Maths Handbook

Liz Heesom

Contents

Introduction

Number

Fractions and decimals

Handling data

Shape and space

Measurement

Skills builder

Introduction

What is an 11+ maths exam?

A maths test is a common test for the 11+ exam and although the content is rooted in the National Curriculum at Key Stage 2, the depth and breadth of questions asked might be more challenging for some children. This may be especially so with children who lack confidence in the subject.

The 11+ exam is taken by children at the beginning of Year 6. It is a test used by state-funded grammar schools or by selective schools for Year 7 onwards. It is used to select the children who perform the best under exam conditions and to place them in a school environment with peers of a similar academic ability. Unlike most other exams, selective entrance tests cannot be retaken. There is no second chance at the 11+ (although some schools do still set exams for entry at 12+ or 13+), so there is often fierce competition to perform well and achieve good results.

There are two main exam boards involved in producing 11+ maths exams; Granada Learning (GL) Assessment and the Centre for Evaluation and Monitoring, Durham University (CEM). GL Assessment uses a separate paper for maths, while CEM tends to mix other 11+ subjects such as English, verbal and non-verbal reasoning with maths to create a paper divided into sections. There are other exam boards and individual schools who write their own papers, and some schools will have the 11+ exam completed on a computer rather than on paper.

An 11+ maths paper can be written in two formats, following either a multiple-choice or standard layout. For a multiple-choice paper, children will need to choose their answer from a set of options and mark it on a separate answer sheet. Answers must be marked in these booklets very carefully as the answer sheets are often read and marked by a computerised system. In the standard format, children must write each answer directly onto the question paper.

As with most exams, 11+ exam papers are timed, typically lasting between 45 minutes and one hour. The introduction of a time-limit can potentially have an impact on a child's performance, so it is important for children to work through practice materials in both timed and non-timed environments.

The scope and content of an 11+ maths test can often differ across UK regions, as there is a range of question types that can be included. However, a paper will generally be intended to test a child's ability to:

- think and calculate quickly
- apply logical thinking and problem solving
- apply their knowledge of times tables

- apply the four number operations (+ − × ÷) accurately
- understand number relationships, measurement, mental arithmetic, geometry and data handling
- work systematically.

These skills are tested through a series of questions that include the following types.

Number

These questions test a child's ability to understand place value, the decimal system, rounding numbers and using the four main operations, including long multiplication and long division techniques. They might cover factors, multiples, problem solving, sequences and algebra.

Fractions and decimals

These questions test a child's ability to understand of fractions of numbers, mixed numbers, improper fractions, equivalent fractions and decimal fractions. Children might be required to add, subtract, divide and multiply fractions, as well as calculate percentages, ratio and proportion.

Handling data

These questions test a child's ability to organise and compare information using a variety of charts, grids, graphs, Venn diagrams and timetables. Children need to be able to calculate the mean, median, mode, range and probability.

Shape and space

These questions test a child's ability to understand 2D and 3D shapes, angles, bearings, area, perimeter, volume and capacity. They might involve triangle facts, quadrilateral and polygon facts and transformations (coordinates, reflection, symmetry, rotation and translation).

Measurement

These questions test a child's ability to understand metric and imperial units of measurement, reading scales, time and timetables. They might cover time and date facts, map scales and problem-solving using length, mass, volume and capacity.

This book will help you to understand the key questions found in 11+ maths exams. The Bond range of maths assessment papers and the CEM maths and non-verbal reasoning books can be used alongside this book. Bond also provides a range of exam-style practice papers in both multiple-choice and standard format.

How to use this book

The book has been divided into sections. How you work through the book is up to you. You can choose to work through the sections in order and complete the questions within them sequentially. Or if you prefer, you can choose any section to start and pick and choose a particular question type within that section.

Each numbered section includes an explanation of the particular question type and the skills that you will need to be able to answer the questions. Make sure you read through all the information carefully before attempting any questions and ask a parent or helper if you do not understand.

Example boxes

For each question type, one or more examples are provided to show you the suggested way to work through the questions and to aid your understanding.

Example boxes look like this:

Have a Go questions

These questions give you the opportunity to check your understanding of what you have just learned. For some of the longer questions, you might find it easier to do your working out on some spare paper.

In each section you will also find useful **Exam Tips** and important information that you should try to **remember** and apply when working through the questions both in practice sessions and in the actual test itself.

We hope you enjoy using the book. Good luck!

Number

1 *Place value*

Knowing the value of digits

You need to know the **value** of a digit wherever it appears in a number.

Multiplying and dividing by 10s, 100s and 1000s

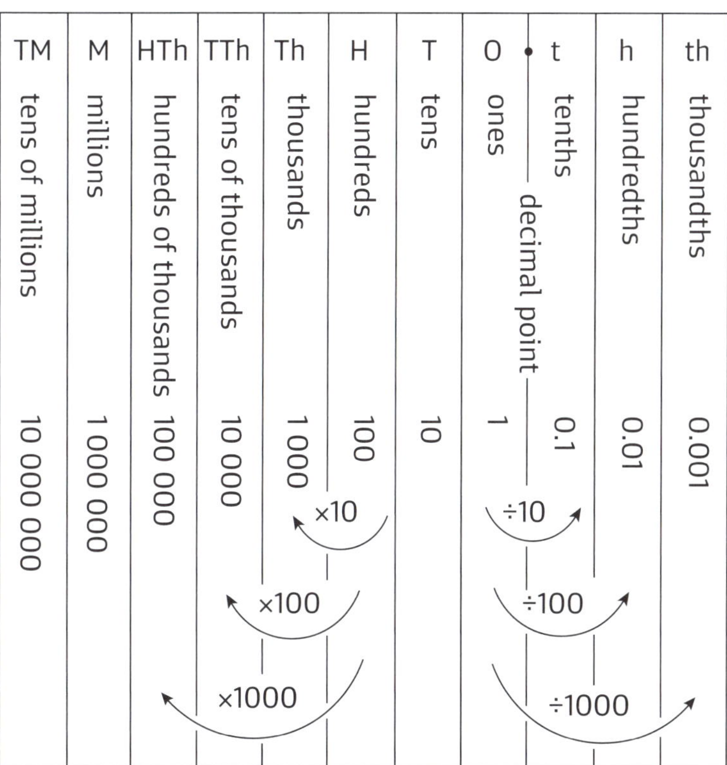

REMEMBER: The 'ones' column is sometimes called the 'units' column.

When a **whole number** is multiplied by:

- **10**, all **digits** move **one** column to the **left**. If the ones column is empty, a zero needs to be written in it.
- **100**, all digits move **two** columns to the **left**. If the ones or tens columns are empty, zeros need to be written in them.
- **1000**, all digits move **three** columns to the **left**. If the ones, tens or hundreds columns are empty, zeros need to be written in them.

Number

When a number is divided by:

- **10**, all digits move **one** column to the **right**.
- **100**, all digits move **two** columns to the **right**.
- **1000**, all digits move **three** columns to the **right**.

"What about decimal numbers?"

As with **whole numbers**, when you use **decimal numbers** all digits in the number have to move in one direction and the same number of columns. Just remember to keep the digits in the same order!

147.5 ÷ 100 = 1.475

"What about numbers with no decimal point?"

Whole numbers can be written with a decimal point but must have one or more zeros in the columns representing tenths, hundredths and so on:

73 = 73.0 = 73.000 000 000…

Rounding numbers

- Rounding a number to the nearest 10 means finding the nearest **multiple** of 10.
- Rounding a number to the nearest 100, 1000 or 10 000 means finding the nearest multiple of 100, 1000 or 10 000 to that number. You can round up or round down.
- Rounding up means finding the next largest multiple of your rounding number.
- Rounding down means finding the next smallest multiple of your rounding number.

So, if you are rounding to the nearest 10, you go up or down to the next multiple of 10.

Look at this example:

Round 12 194 to the nearest 100.

12 194

The digit in the tens column is **greater than 5** so round up to the next multiple of 100.

194 is closer to 200 than 100.

12 194 **rounded to** to the nearest 100 is **12 200**.

LEARN: Look at the number in the column to the right of the one you are asked to round to. If it is less than 5, round down. If it is 5 or greater than 5, round up.

The same idea works for decimals, so you can round to a certain number of decimal places. To decide whether you must round up or down, look at the digit to the right of the decimal place to which you need to round.

For example:

- 12.264 rounded to one decimal place (the nearest **tenth**) is 12.3
- 12.264 rounded to two decimal places (the nearest **hundredth**) is 12.26

Number

KEY FACTS

- The digits to the left of the decimal point show the number of ones, tens, hundreds, thousands and so on. The digits to the right of the decimal point show the number of tenths, hundredths, **thousandths** and so on.
- To multiply by 10, 100 or 1000, imagine the digits are written in the place value columns and move them to the left to make the number larger.
- To divide by 10, 100 or 1000, move the digits to the right to make the number smaller.
- If the digit in the column to the right of the one you are asked to round to is less than 5, round down. If it is 5 or greater than 5, round up.

HAVE A GO

1 Round 345 678 to the nearest:

a 10 ______________________ **b** 100 ______________________

c 1000 ______________________ **d** 10 000 ______________________

2 What are the values of the 3 marked with x and the 3 marked with y in this number?

234.93

↑ ↑

x y

x ______________________

y ______________________

3 Write 1.55 to the nearest tenth. ______________________

Number

4 Divide 305 by 1000. ______

5 Round 476.528 to the nearest:

a tenth ______ b hundredth ______

EXAM TIP

When working with place value and rounding, jot down the column headings to remind yourself of the order, like this:

TM M HTh TTh Th H T O/U · t h th

Practise doing this before the exam so that you are sure of the order of places.

2 Addition and subtraction problems

"I never know whether I am supposed to be adding or subtracting."

Addition

An easy way to identify problems that can be solved using addition is by looking out for these words, phrases and symbols – they all mean 'add':

+	plus	increase
how many altogether	add	find the total
find the sum		

Adding a whole number to another number gives an answer larger than the number you started with.

To add in your head, it can be helpful to start from the larger number and add on the smaller number or numbers in easy jumps.

Remember the jumps as you go and then add the jumps together.

So, to add 329 and 1572, start with 1572. Add 300, then 20, then 9.

1572 + 300 = 1872

1872 + 20 = 1892

1892 + 9 = 1901

It can be helpful to write the numbers one under the other to ensure you add together the digits in their columns.

For example:

1 3 2 2	
2 4 8 3	
+ 5 2 1 9	
8 0 0 0	Total the thousands
9 0 0	Total the hundreds
1 1 0	Total the tens
1 4	Total the ones
9 0 2 4	The grand total

Subtraction

An easy way to identify problems that can be solved using subtraction is by looking out for these words, phrases and symbols – they all mean 'subtract':

–	decrease	reduce
subtract	find the difference	take away
minus	deduct	

Subtracting a **positive number** from another number gives an answer smaller than the number you started with.

To subtract in your head, it can be helpful to start from the smaller number and count up to the larger number in easy jumps. Remember the jumps as you go and then add the jumps together.

Finding the difference between two numbers means the smaller number must be subtracted from the larger number.

Find the difference between 843 and 387.

843 – 387

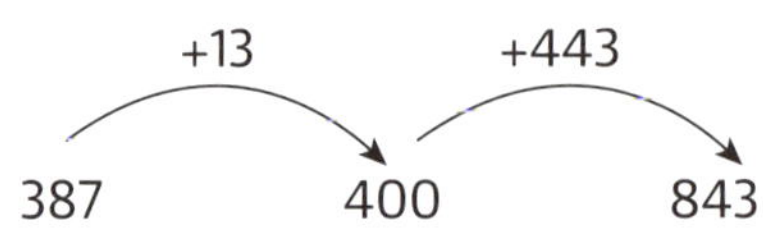

443 + 13 = 456

The difference between 843 and 387 is 456.

Addition and subtraction involving money

Rounding up to the next pound can make calculations involving money much easier to work out in your head. For example:

99p is nearly £1 | £1.99 is nearly £2 | £7.99 is nearly £8

Do not forget to deal with the penny difference and take it off your answer!

KEY FACTS

- The opposite or **inverse** of addition is subtraction.
- The opposite or inverse of subtraction is addition.
- Add is the same as find the total, plus, increase, find the sum, how many altogether.
- Subtract is the same as take away, minus, find the difference, find the change, decrease, reduce, deduct.
- **Estimate** answers (make a sensible guess) before you work out the actual answer.

HAVE A GO

1 Find the sum of 1786 and 2965. ______________________________

2 Find the difference between 1786 and 2965. ______________________________

3 Increase £7.99 by £2.50. ____

4 Decrease £5.65 by £1.99. ____

5 What is 5001 – 356? ____

EXAM TIP

In an exam, it is vital that you work carefully, neatly and in an orderly way. Use your fingers, quick drawings, jottings or whatever you need to find the correct answer as quickly as possible.

3 Multiplication and division problems

"I am never sure whether to multiply or to divide."

Multiplication

An easy way to identify problems that can be solved using multiplication is by looking out for these words, phrases and symbols – they all mean 'multiply':

× multiply times find the product groups of lots of

Multiplying a number by a positive **integer** (whole number) gives an answer larger than the number you started with.

A calculation using column multiplication can be written quickly and helps to break down difficult or long multiplication questions such as 326 × 42.

1. Set up the multiplication as a grid.

$$\begin{array}{r} 3\ 2\ 6 \\ \times \quad 4\ 2 \\ \hline \\ \hline \end{array}$$

Number

2. Begin with the 2 in the 'ones' or 'units' column:

   2 × 6 = 12 so write the 2 in the ones column answer space and carry over the 1 ten to the tens column. Write it in small numbers above or below the tens column numbers.

   Next, multiply 2 × 2 and add on the 1 you have carried over. This equals 5, so write the 5 in the tens column answer space.

   Finally, you have 2 × 3 = 6 so write the 6 in the hundreds column.

$$\begin{array}{r} {\scriptstyle 1} \\ 3\,2\,6 \\ \times \quad 4\,2 \\ \hline \mathbf{6\,5\,2} \\ \hline \end{array}$$

   326 × 2 = 652

3. Now you need to multiply by the 4 in the tens column, which is really multiplying by 40.

   As 40 = 4 × 10, you can put the 0 in the ones column to hold the place and then multiply each number by 4. That is the same as multiplying by 40.

$$\begin{array}{r} {\scriptstyle 1\;2\;1} \\ 3\,2\,6 \\ \times \quad 4\,2 \\ \hline 6\,5\,2 \\ \mathbf{1\,3\,0\,4\,0} \\ \hline \end{array}$$

   326 × 40 = 13 040

   Calculate 4 × 6 = 24 and write the 4 in the tens column. Carry over the 2 and write it above or below the hundreds column.

   Now work out 4 × 2 and add on the 2 you have carried over. This equals 10, so write 0 in the hundreds column. Carry over the 1 and write it above or below the thousands column.

   Next, work out 4 × 3 and add on the 1 you have carried over. This equals 13 so write the 13.

4. To complete this multiplication, add up the two answers you have found.

```
  1 2 1
    3 2 6
  ×   4 2
  -------
    6 5 2
1 3 0 4 0
---------
1 3 6 9 2
```

652 + 13 040 = 13 692

326 × 42 = 13 692

Division

An easy way to identify problems that can be solved using division is to look out for these words, phrases and symbols – they all mean 'divide':

÷ $\overline{)}$ share   split into equal amounts   /   how many … in …   find the **fraction**

Dividing a number by a positive integer (whole number) gives an answer smaller than the number you started with.

A number is **divisible** by a smaller number if the smaller number (or **divisor**) divides exactly into the larger number (or **dividend**). 25 is divisible by 5 because 5 divides into 25 exactly. The answer is the **quotient**.

REMEMBER: The number you get when you divide one number, a dividend, by another, a divisor, is called a quotient.

This list of quick ways to test whether a number is divisible is very useful to learn.

A number is divisible by or a multiple of:

- 2 if it is even
- 3 if the sum of the digits is divisible by 3
- 4 if the last two digits are a number divisible by 4
- 5 if the ones digit is 5 or 0
- 6 if the number is even and divisible by 3
- 9 if the sum of the digits is divisible by 9
- 10 if the ones digit is 0

See *Topic 5: Factors and multiples* to learn about multiples.

You can divide by using repeated subtraction. This means taking the same amount away again and again until there is nothing left, or the **remainder** is smaller than the number you are subtracting. Your answer is the number of times you were able to subtract the amount, with the remainder if there is one.

REMEMBER: The remainder is sometimes very important. If you have a remainder when solving a division problem, read the question very carefully to see if you need to include the remainder in your answer somehow.

Look at this example:

How many pencils costing 30p each can be bought for £8?

1. Set up the 'bus stop method' for saying "How many 30ps are in £8?".
   Change £8 to 800p so you are working using pence, not a mixture of £s and ps.

```
   _____
30)8 0 0
```

2. How many 30s fit into 8?
   None, because 8 is smaller than 30, so place a '0' above the 8 and look at 8 and the next digit together, 80.

```
   0
   _____
30)8 0 0
```

3. How many 30s fit into 80?
   2 × 30 = 60 so write '2' above the first '0' and place 60 underneath the 80.
   Then subtract 60 from 80 to give 20.

```
   0 2
   _____
30)8 0 0
   6 0
   ---
   2 0
```

4. Next, bring down the last '0' from the dividend number to give 200.

```
   0 2
   _____
30)8 0 0
   6 0↓
   ---
   2 0 0
```

5. How many 30s fit into 200?
   6 × 30 = 180 so write '6' above the final '0' in the dividend and place 180 underneath the 200.
   Subtract 180 from 200 to give a remainder.

```
   0 2 6
   _____
30)8 0 0
   6 0↓
   -----
   2 0 0
   1 8 0
   -----
     2 0
```

6. The remainder is 20p, not enough money to buy another pencil, so the final answer is 26 pencils.

26 pencils costing 30p each can be bought for £8.00 with 20p left over.

Number

KEY FACTS

- The opposite or inverse of multiplication is division.
- The opposite or inverse of division is multiplication.
- Multiply is the same as find the product, times, groups or lots of.
- Divide is the same as share or split into equal amounts, partition or find a fraction. You can say 'How many … in …?' to help your thinking.
- A number is divisible by a smaller number if the smaller number (divisor) divides exactly into the larger number (dividend), with no remainder.
- The answer in a division calculation is the quotient.

HAVE A GO

1 Find the cost of 6 books that each cost £6.95. ______________________

2 The product of two numbers is 168. If one of the numbers is 24, what is the other number? ______________________

3 Circle the quotient for this division calculation:

6300 ÷ 70 = 90 19 29 9 900

4 What is the smallest number divisible by 2, 3 and 4? ______________________

5 Multiply 630 by 55. ______________________

Number

4 Mixed or several-step problems

"I never know what to do first with problems."

Problem-solving steps

Problems are often made up of many stages and you will need several different operations in order to find the answers. Following a clear set of steps, such as the '6-point plan' below, could make problems easier to work out.

1. Read the problem carefully. Think: "What am I being asked to work out?" Highlight or underline key words.

2. Decide what you have to work out and which number operation you need. Write the calculation down.

3. Continue deciding what you must work out at each step and which number operation you need. Write the calculations.

4. **Estimate** the answer.

5. Work out the answers to the calculations in order.

6. Ask yourself: "Is my answer reasonable compared with my estimate?"

Another method is the 'RUCSAC' method:

Read, **U**nderstand, **C**hoose, **S**olve, **A**nswer, **C**heck

KEY FACTS

- There are four number operations: +, −, × and ÷. Look again at Topics 2 and 3 to remind yourself of the different mathematical words used for these signs.
- Try following the 6-point plan to help decide what you are being asked to do in a several-step problem.
- Check how many points the question is worth so you can make sure your answer has enough detail.
- Profit means how much extra money you can earn if you sell something for more than it cost.

HAVE A GO

1 Rosie bought 5 m of curtain tape. She gave the shop assistant £20 and received £1.75 in change. How much did the tape cost per metre? __________

2 A village school of 57 children has one class of 11 children. The rest are divided into two equal classes. How many children are there in each of these two classes? __________

3 How much change from £10 does Sharee have if she buys two notebooks for £2.40 each and a pen for £2.95? __________

4 Ed bought 14 toys for £1.35 each and then sold each toy for £1.70. How much was his total profit? __________

5 Samir has 170 stickers. He puts 18 stickers on each of 9 pages in his sticker album. How many stickers does he have left over? __________

EXAM TIP

Some of the questions in an 11+ exam are bound to be questions like this. Try to learn and practise the 6-point plan and, in an exam, really imagine what is going on in a problem. If it helps, draw pictures or imagine yourself as the person mentioned in the problem, if there is one. These kinds of questions may give more than one point, so make sure to find out how many points are given for each question. If you can, show the steps you take to work out your answer to gain all the possible points and to prove how you got your answer.

5 Factors and multiples

Factors

A factor is a whole number that divides exactly into another number. An easy way to be sure of finding all the factors in a number is to find them in pairs like this:

What are the factors of 48?

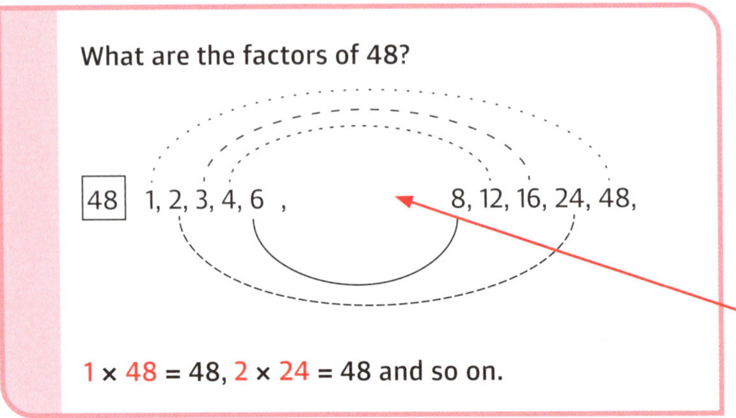

1 × 48 = 48, 2 × 24 = 48 and so on.

You know you have found all the factors when there are no possible factors between the middle two numbers.

You may have to find the **highest common factor (HCF)** of a set of numbers. To do this, find all the factors of the numbers you are given and see which factors they have in common. The largest one is the highest common factor.

Look at this example:

Find the HCF of 24, 42 and 48.

The factors of 24 are: 1, 2, 3, 4, 6, 8, 12, 24.

The factors of 42 are: 1, 2, 3, 6, 7, 14, 21, 42.

The factors of 48 are: 1, 2, 3, 4, 6, 8, 12, 16, 24, 48.

The HCF of 24, 42 and 48 is 6.

REMEMBER: A factor is a whole number that will divide exactly into another number.

See *Topic 6: Special numbers* to remind yourself about **prime numbers**.

Multiples

A multiple of a number is the answer when it is multiplied by another number.

$4 \times 5 = 20$ 20 is a multiple of 4 and 5.

The multiples of 5 are the numbers when you count in fives: 5, 10, 15, 20, 25 and so on.

You may have to find the **lowest common multiple (LCM)** of a set of numbers. To do this, write down the first few multiples of all the numbers, starting with the highest number in the set. Check to see if one of these is a multiple of all the numbers. If not, try a few more until you find the LCM.

Look at this example:

LEARN: Prime factors are factors which are also prime numbers.

REMEMBER: You have to multiply to get a multiple.

Find the LCM of 4, 6 and 9.

The first five multiples of 9: 9, 18, 27, 36, 45

The first nine multiples of 4: 4, 8, 12, 16, 20, 24, 28, 32, 36

The first six multiples of 6: 6, 12, 18, 24, 30, 36

The LCM of 4, 6 and 9 is 36.

HAVE A GO

1 Find the lowest common multiple of 10, 12 and 15. ______

2 Circle the prime factors of 77:   6   7   8   9   10   11

3 Write down all the factors of 72. ______

4 What is the smallest number that is a multiple of 3, 5 and 6? ______

5 What is the highest common factor of 32, 88 and 120? ______

Number

KEY FACTS

- A factor is a whole number that will divide exactly into another number.
- The HCF (highest common factor) of a set of numbers is the largest number that is a factor of all numbers in the set.
- The **prime factors** of a number are the prime numbers which can be multiplied together to make that number.
- A multiple of a number is the answer when the number is multiplied by another number. If you count in 12s, the numbers you say are the multiples of 12.
- The LCM (lowest common multiple) of a set of numbers is the smallest number that is a multiple of all numbers in the set.

6 Special numbers

There are some particular types of numbers that you need to be able to recognise for the 11+. A short section on each of the eight main types is given below.

Negative numbers

On a number line, **positive numbers** are to the right of 0 (zero) and are greater than 0.

Negative numbers are to the left of 0 and are less than 0.

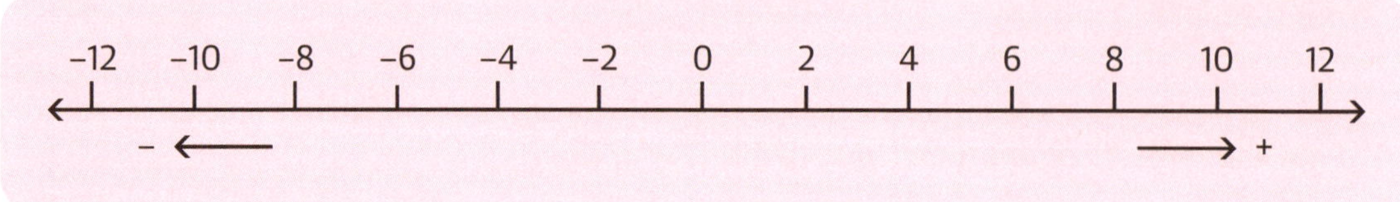

A thermometer shows positive and negative numbers on a **vertical** scale. Freezing point is zero **degrees** Celsius (0°C). Above zero are the positive numbers (getting warmer). Below zero are the negative numbers (getting colder).

Square numbers

When you multiply a whole number by itself, you get a **square number**. You can see why they are called this by looking at the diagram on the next page.

REMEMBER: A square number is a number multiplied by itself.

The first twelve square numbers are:

1, 4, 9, 16, 25, 36, 49, 64, 81, 100, 121, 144.

4^2 means 'four squared', or four times four. The number in little writing (the **index** number) tells you that you must multiply 4 by itself.

$1 \times 1 = 1^2 = 1$

$2 \times 2 = 2^2 = 4$

$3 \times 3 = 3^2 = 9$

$4 \times 4 = 4^2 = 16\ldots$

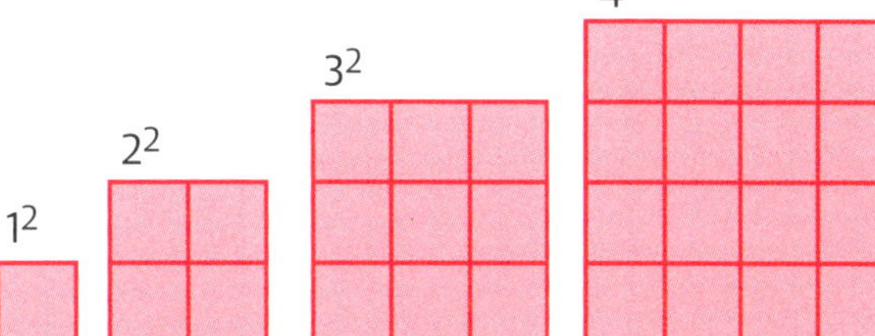

Cube numbers

When you multiply a whole number by itself and then multiply the answer by the same number again, you get a **cube number**. You can see why they are called this by looking at the diagram below.

REMEMBER: A cube number is a number multiplied by itself and by itself again.

The first six cube numbers are:

1, 8, 27, 64, 125, 216.

4^3 means 'four cubed', or four times four times four. The index number tells you that you must multiply 4 by itself then multiply the answer by 4.

$1 \times 1 \times 1 = 1^3 = 1$

$2 \times 2 \times 2 = 2^3 = 8$

$3 \times 3 \times 3 = 3^3 = 27$

$4 \times 4 \times 4 = 4^3 = 64\ \ldots$

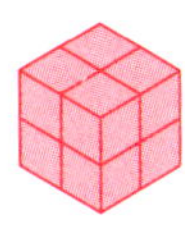

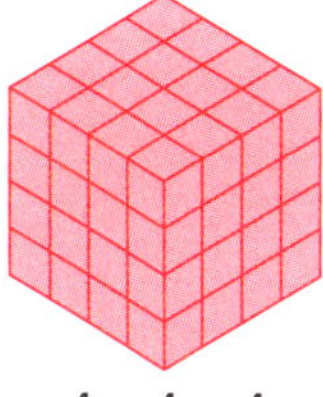

$1 \times 1 \times 1$ $2 \times 2 \times 2$ $3 \times 3 \times 3$ $4 \times 4 \times 4$

When a **cube** is built using **centimetre** cubes, starting with just one cube for the first cube, the number of cubes used to make the increasingly large cubes is the same as the cubed numbers.

Consecutive numbers

Consecutive numbers are numbers that follow on in order. So numbers such as 45, 46, 47, 48, 49, 50 are consecutive.

The numbers in a group like 36, 45, 47, 50, 49 are not consecutive.

LEARN: Consecutive means 'in a row', or 'following in order'.

Prime numbers

A prime number can only be divided exactly by two numbers, 1 and the number itself. It has only two factors.

The first 10 prime numbers are all less than 30:

2, 3, 5, 7, 11, 13, 17, 19, 23, 29

1 is not a prime number because it has only one factor: 1

2 is the only even prime number. Can you see why?

LEARN: A prime number has only two factors: 1 and the number itself.

Roman numerals

Here are seven **Roman numerals** and their standard number **values**, the ones mainly used now:

Roman	I	V	X	L	C	D	M
Standard	1	5	10	50	100	500	2000

You can make other whole numbers by using a combination of these seven Roman numerals. For example:

IV = 4 (one less than 5)
VI = 6 (one more than 5)
IX = 9 (one less than 10)
XIII = 13 (10 + 3)
DXI = 511 (500 + 10 + 1)
CM = 900 (100 less than 1000)

LEARN: Dates and kings/queens are sometimes written using Roman numerals:

MMXV = 2015
Henry VIII

You may also need to know the value of these Roman numerals:

XL = 40 LX = 60 XC = 90

Triangular numbers

If you arrange dots in a triangular pattern, then the increasing number of dots needed to make a triangle form the sequence of **triangular numbers**.

These are the first 10 triangular numbers:

1, 3, 6, 10, 15, 21, 28, 36, 45, 55

The difference between each pair of numbers goes up by 1 each time. The first difference is 2, then 3, then 4 and so on.

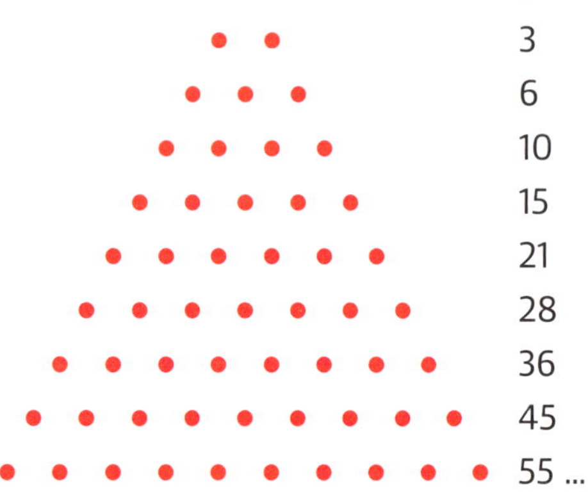

Square root

The **square root** of a number is the number you multiply by itself to make that number. This is the opposite of finding a square number.

The square root of 100 is 10 because 10 × 10 is 100.

The square root of 49 is 7 because 7 × 7 is 49.

The symbol for a square root is: √. So, 'the square root of 49' is written as √49.

Number

HAVE A GO

1 $3^2 + 5^2 =$ ______________

2 Which triangular number is the same as the square root of 81, added to 2^2, added to the prime number between 20 and 28? ______________

3 Find three consecutive numbers that add up to 42. ______________

4 Circle the prime numbers: 40 41 42 43 44 45 46 47 48 49

5 Subtract √81 from 5^3. ______________

KEY FACTS

- Negative numbers are less than zero: –1, –2, –3 ... –74, ... –201, ...
- Square numbers are the result of multiplying a number by itself: 1 × 1, 2 × 2, … 6 × 6, …
- Cube numbers are the result of multiplying a number by itself and by itself again: 1 × 1 × 1, 2 × 2 × 2, ... 6 × 6 × 6, …
- Consecutive numbers follow on in order: for example, 21, 22, 23.
- Prime numbers have only two factors, 1 and the number itself: 2, 3, 5, 7, … So 1 is not prime as it has only one factor.
- Roman numerals use letters to represent numbers: for example, V = 5, X = 10, L = 50, C = 100.
- Triangular numbers start at 1, then add 2, then add 3 and so on: 1, 3, 6, 10, 15, …
- A square root is the number you multiply by itself to make a square number.

7 Sequences

Common number patterns

Maths is largely about seeing patterns. You need to know and be able to recognise some of the most common and useful patterns. Here are some sequences and patterns you should already know.

Odd numbers:	1, 3, 5, 7...
Even numbers:	2, 4, 6, 8 ...
Multiples:	for example, 7, 14, 21, 28 ...
Prime numbers:	2, 3, 5, 7...
Square numbers:	1, 4, 9, 16 ...
Cube numbers:	1, 8, 27, 64 ...
Triangular numbers:	1, 3, 6, 10 ...

You will also find doubling and halving numbers useful.

Doubling numbers:	1, 2, 4, 8 ...
Halving numbers:	64, 32, 16, 8 ...

Finding the rule

To find the missing number in a sequence, first find out the rule. If all the numbers are getting bigger or smaller, it can be very helpful to write the difference between each pair of numbers along the sequence to help you find the rule.

Some sequences will go up or down by the same number each step. Others may go up or down one more or less than the previous step.

Be careful: some patterns go alternately in pairs or even in threes! So, if you cannot find a pattern between pairs of numbers and the numbers are not in size order, check to see if there are two or three sequences in one.

For example:

Write the next two numbers in this sequence:

1, 2, 3, 4, 6, 6, 10, 8, 15, ______ , ______

Is there a pattern?
No, the numbers look random.

Try looking at every other number (alternate numbers):

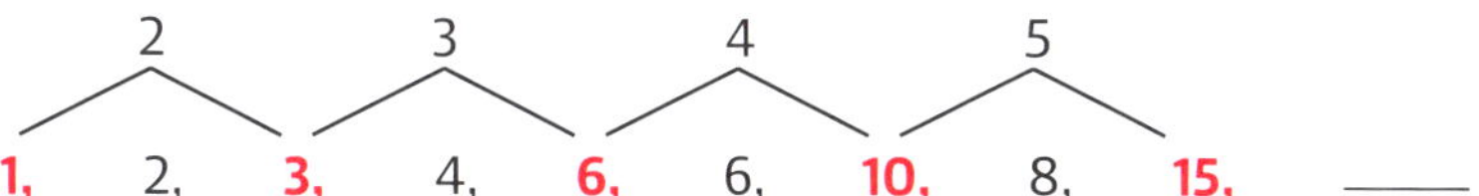

1, 2, **3,** 4, **6,** 6, **10,** 8, **15,** ____

These are triangular numbers. The next number would be 21 (add 6).

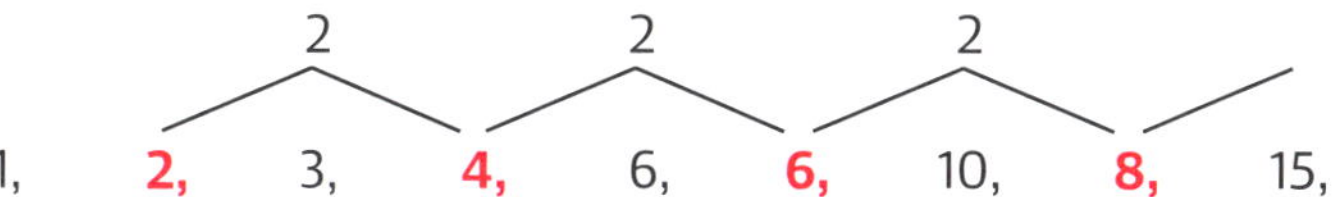

1, **2,** 3, **4,** 6, **6,** 10, **8,** 15,

These are even numbers. The next number would be 10.

The next two numbers in the sequence are 10 and 21.

EXAM TIP

In an 11+ exam, there may be questions like the example above. There may be other sequences where you have to find missing numbers not at the end, but in the middle or at the beginning. These are still following patterns and you have to work out what clues the jumps between the numbers give you. Similar puzzles often come up in verbal reasoning 11+ exams too.

HAVE A GO

Find the missing number in each of these sequences:

1 47, 52, 58, 65, 73, ____

2 100, 90, ____ , 73, 66, 60

3 4, 8, ____ , 32, 64, 128, 256

4 ____ , 3, 6, 10, 15, 21, 28

5 1, 2, 3, 4, 5, 8, 7, 16, 9, ____

KEY FACTS

- To find the missing number in a sequence, first find out the rule.
- Some sequences will go up or down by the same number each step. Others may go up or down one more or less than the previous step.
- Some sequences have more than one rule.

Number

8 Equations and algebra

REMEMBER: Notice that 'equation' comes from the word 'equal', meaning 'the same' or 'identical'.

Equations

An **equation** is a number sentence where one thing is equal to something else.

An equals sign (=) shows that the number (or the answer to the calculation) on the left of it must be equal to, or have the same value as, the number (or the answer to the calculation) on the right of it.

Look at this equation: 3 + 4 = 7

Three plus four is equal to seven. The answer to the calculation on the left of the equals sign has the same value as the number on the right of it. Both sides of the equation are balanced. They are both equal to seven. It can be called a simple equation because it has an equals sign.

Here is another equation:

$\_\_\_\_ \times 6 = 2 \times 9$

You can only work out 2×9 on the right of the equals sign, but you know the answer will be the same on both sides, so the missing number must be 3.

$3 \times 6 = 2 \times 9$

Algebra

Algebra is about finding unknown (or mystery) numbers in equations. The unknown numbers are often shown as letters such as a, b, c, x, y, z or symbols such as ❄, ■ or ◆. Notice how letters standing for mystery numbers can be written differently so that, for instance, mystery number x is different from the multiply sign ×.

You have to find the value of the unknown number to solve an equation.

For example:

> Solve $x = 8 - 3$.
>
> x must be equal to the value of the numbers on the other side of the equals sign. 8 – 3 is 5. So, $x = 5$.

Sometimes you might see a number next to a letter, for example: $2y$.

It is a short way of saying $2 \times y$.

You might also see, for example, $\frac{10a}{5}$.
It means the same as 10 times a divided by 5, or $10a \times 5$.

Now look at these examples:

If $x = 2$ and $y = 4$, find $\frac{3x}{6y}$.

$\frac{3x}{6y} = \frac{3 \times 2}{6 \times 4} = \frac{6}{24} = \frac{1}{4}$

So $\frac{3x}{6y} = \frac{1}{4}$

If $x = 10$ and $y = 5$, find $\frac{3x}{6y}$.

$\frac{3x}{6y} = \frac{3 \times 10}{6 \times 5} = \frac{30}{30} = 1$

So $\frac{3x}{6y} = 1$

When finding an unknown number, it can be helpful to use inverse operations so that the unknown number is left on its own on one side of the equation. To do this, look at the sign and do the opposite to both sides of the equation.

For add → subtract.
For subtract → add.
For multiply → divide.
For divide → multiply.
For double → halve.
For halve → double.

Look at these examples:

$x - 3 = 9$ What does x equal?

To leave x on its own, add 3 to both sides of the equation to remove the '– 3':

$x - 3 + 3 = 9 + 3$

$x = 9 + 3$

$x = 12$

$\frac{a}{5} = 4$ What does a equal?

To leave a on its own, multiply both sides of the equation by 5 to remove the '× 5'.

$\frac{a}{5} = 4$

$\frac{a}{5} \times 5 = 4 \times 5$

$a = 20$

Number

You may have to solve equations that need more than one inverse operation before the mystery number is left on its own on one side of the equation.

Look at this example:

$10x - 2 = 7x + 4$ What does x equal?

$10x - 2 + 2 = 7x + 4 + 2$

$10x = 7x + 6$

$10x - 7x = 7x + 6 - 7x$

$3x = 6$

$3x \div 3 = 6 \div 3$

$x = 2$

REMEMBER: Indices are index numbers like the little ^{2} in 8 squared or the little ^{3} in 2 cubed.

LEARN: Always use BIDMAS. Work out the Brackets first, then any Indices, then Division and/or Multiplication, then Addition and/or Subtraction.

Sometimes there are brackets in calculations, for example, $3 + (7 - 2)$.

To complete this, you must first do the calculation in the brackets: $7 - 2 = 5$.

The calculation then becomes $3 + 5 = 8$.

HAVE A GO

1 A number multiplied by itself and doubled is 50. What is the number? ________

2 $4a + 12 = 7a - 9a =$ ________

3 $\frac{z}{9} = 5z =$ ________

4 If $s = 4$, $t = 6$ and $u = 9$, solve $(s + t^2) - 9$. ________

5 $7 \times ? = 65 - 9$ Find the value of ? ________

EXAM TIP

Often people get very confused in exams when they see letters standing in for numbers. In 11+ maths, the missing numbers are not likely to be very large, so you can always try out numbers in the gaps and see if they work. If you cannot work out an answer, try saying the equation to yourself using the word 'something' in place of the mystery number. It may make more sense then.

KEY FACTS

- $4z = 4 \times z$
- $\frac{3x}{2} = 3x \div 2$ or 3 times x divided by 2
- BIDMAS shows the order of operations.
- Use inverse operations if the first number is missing in a calculation or to get the mystery number on its own.

9 Function machines

Function machines

A function machine applies one or more rules to a number to give another number.

- A number is fed in from the left, which could be a mystery number. This number is sometimes called the **input**.
- Next, one or more boxes tell you to do something to the number using the number operations (**+**, **×**, **−**, **÷**).
- Then the answer comes out at the right-hand side of the machine. This number is sometimes called the **output**.

If the missing number is to the right of the function machine, apply the rule or rules that you are given, in turn, to the number on the left and find the missing number.

If the missing number is to the left of the function machine, use the opposite (or inverse) of the rule or rules you are given, working backwards to find the missing number.

In this example, the missing number is to the left of the function machine and you therefore need to begin with the answer to the right of the machine.

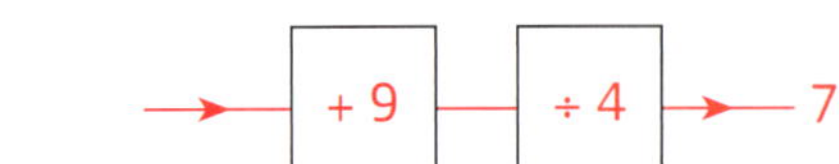

The two number operations need to be carried out from right to left, using their inverses, so:

The **output** number is 7.

÷ 4 becomes × 4: $7 \times 4 = 28$

+ 9 becomes – 9: $28 - 9 = 19$

The **input** is **19**.

You can check your answer by working through the function machine starting off with the answer you found:

19 + 9 = 28 ⟶ 28 ÷ 4 = 7 ✓

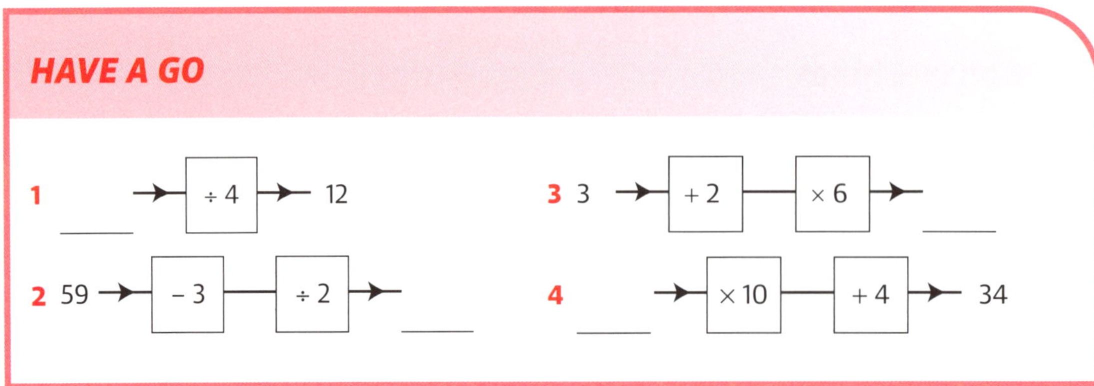

KEY FACTS

- Function machines work from left to right. A number is fed in (input) on the left, number operations are applied to it in turn, and the answer (output) comes out on the right.
- If the input number is missing, start with the output number and apply the opposite (inverse) number operations in order, reading from right to left to find the answer.
- Always double-check your answer when you have worked in reverse by going forwards through the function machine.

Fractions and decimals

Fractions

Writing a fraction

"I find fractions confusing."

Learning about **fractions** means understanding some long words. Having pictures in your mind for each new word will help to sort out any confusion.

There are two parts to a simple fraction:

$\frac{1}{2}$ 1 ← the top part is called the numerator

2 ← the bottom part is called the denominator

The **denominator** tells you how many equal parts the 'something' has been divided or cut up into. It also gives the fraction its name. Two equal parts are halves, three equal parts are thirds, four equal parts are quarters, five equal parts are fifths and so on.

The **numerator** tells you how many of the equal parts you are looking at.

Fractions like these are called **common fractions**.

It can be useful to draw quick pictures of fractions.

For example, to draw diagrams to represent $\frac{1}{8}$, draw one whole shape. You can draw a circle or a box. Divide the shape into 8 equal pieces. Colour in one of the equal pieces. Then it is easy to show $\frac{3}{8}$ or $\frac{7}{8}$.

LEARN: A fraction is a part of one whole something.

The 'something' might be one pizza cut into 12 equal slices or a box of 28 pencils shared equally among four children.

Denominator sounds a bit like name.

Numerator is a word like *number*.

LEARN: When you look at a fraction, you can read the line as 'divided by' or 'out of', like a score. So $\frac{3}{8}$ is the same as three divided by eight, or three out of eight.

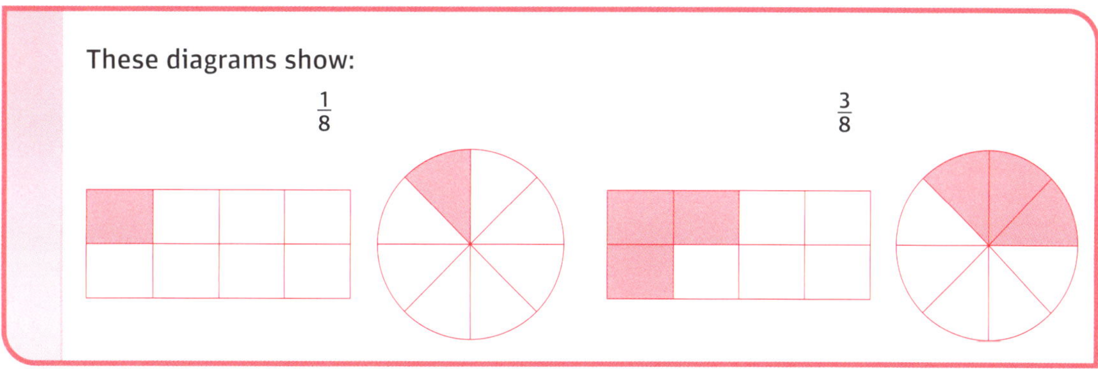

Fractions and decimals

Fractions of numbers

To find a fraction of a number, think about how many equal parts you are dividing the whole amount into (denominator) and how many of those parts you need (numerator).

Divide the number by the denominator and then multiply the answer by the numerator.

For example:

LEARN: Use your times tables knowledge when working out fractions.

Find $\frac{1}{6}$ of 24.		Find $\frac{5}{6}$ of 24.
$24 \div 6 = 4$	← divide by the denominator →	$24 \div 6 = 4$
$4 \times 1 = 4$	← multiply by the numerator →	$4 \times 5 = 20$
$\frac{1}{6}$ of 24 = 4		$\frac{5}{6}$ of 24 = 20

HAVE A GO

Work out the following:

1 $\frac{1}{3}$ of 27 = ______

2 $\frac{4}{7}$ of 63 = ______

3 $\frac{3}{8}$ of 24 = ______

4 $\frac{5}{6}$ of 72 = ______

5 $\frac{7}{12}$ of 48 = ______

Mixed numbers

A **mixed number** is a mixture of a **whole number** and a fraction, like $1\frac{1}{3}$ or $3\frac{3}{5}$.

You might be asked to draw diagrams to show mixed numbers:

These diagrams show:

$1\frac{1}{3}$ $3\frac{3}{5}$

Practise doing this for other mixed numbers you come across, like $2\frac{1}{2}$ or $7\frac{5}{8}$ or $9\frac{3}{4}$.

Improper fractions

Improper fractions are 'top-heavy' fractions. The numerator is larger than the denominator:

$\frac{13}{5}$ $\frac{7}{6}$ $\frac{97}{4}$

Improper fractions can be changed into mixed numbers using division.

Divide the numerator by the denominator.

The answer is the whole number.

The **remainder** becomes the numerator of the fraction. You must keep the denominator the same.

For example:

Change $\frac{13}{5}$ into a mixed number.

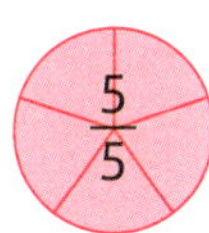

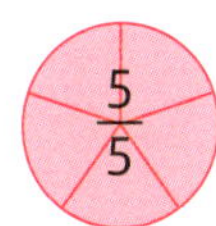

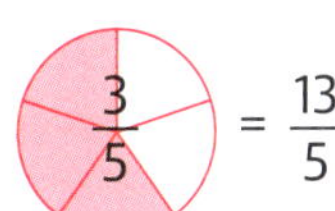

$13 \div 5 = ?$ How many 5s in 13?

$13 \div 5 = 2$ remainder 3

$\frac{13}{5} = 2\frac{3}{5}$ The denominator stays the same.

HAVE A GO

Change these improper fractions into mixed numbers.

1 $\frac{43}{5}$ = ______________________ **2** $\frac{77}{6}$ = ______________________

3 $\frac{91}{10}$ = ______________________ **4** $\frac{63}{4}$ = ______________________

5 $\frac{113}{12}$ = ______________________

Equivalent fractions

Equivalent fractions are fractions that are equal to each other. For example, two quarters are exactly the same as one half:

$\frac{2}{4} = \frac{1}{2}$

To find a fraction equivalent to another fraction, multiply the numerator and the denominator by the same number.

For example, $\frac{7}{8}$ is the same as $\frac{14}{16}$ $\frac{21}{24}$ $\frac{28}{32}$ $\frac{35}{40}$

Can you see why?

Look at this multiplication grid, which shows the same equivalent fractions.

×	2	3	4	5
7	14	21	28	35
8	16	24	32	40

It can be helpful to draw diagrams.

LEARN: Multiplying the numerator and denominator by the same number gives you an equivalent fraction.

This diagram compares thirds and sixths.

It is now easy to see that $\frac{1}{3}$ is equal to $\frac{2}{6}$.

HAVE A GO

Match up five pairs of equivalent fractions.

$\frac{2}{3}$ $\frac{3}{8}$ $\frac{35}{45}$ $\frac{1}{4}$ $\frac{16}{24}$ $\frac{21}{84}$ $\frac{20}{48}$ $\frac{21}{56}$ $\frac{5}{12}$ $\frac{7}{9}$

Decimal fractions

See *Topic 11: Decimal numbers*.

Comparing and ordering fractions

To do this, you may need to look at fractions and find a **common denominator.**

If you have thirds and sixths, the denominators are 3 and 6.
They share the **lowest common multiple** 6. 6 is their common denominator. $\frac{1}{3} = \frac{2}{6}$.

If you have thirds and quarters, the denominators are 3 and 4. They share the lowest common multiple 12. 12 is their common denominator.

$\frac{1}{3} = \frac{4}{12}$ $\frac{1}{4} = \frac{3}{12}$

Now you can see that $\frac{1}{3}$ is larger than $\frac{1}{4}$.

You can use quick drawings and what you know about equivalent fractions to find the largest or smallest fraction and then order them. For example:

Put the following numbers in order of size from smallest to largest.

$\frac{21}{3}$ $\frac{11}{2}$ $\frac{3}{8}$ $\frac{21}{4}$ $\frac{4}{5}$

1. $\frac{21}{3}$ and $\frac{21}{4}$ will be the largest. $\frac{1}{3}$ is larger than $\frac{1}{4}$.
2. Next comes $\frac{11}{2}$.
3. Find a common denominator for $\frac{3}{8}$ and $\frac{4}{5}$.
   It is 40.
   $\frac{3}{8} = \frac{15}{40}$ and $\frac{4}{5} = \frac{32}{40}$, so $\frac{4}{5}$ is larger than $\frac{3}{8}$.

Now you can write the fractions in order of size from smallest to largest:

$\frac{3}{8}$ $\frac{4}{5}$ $\frac{11}{2}$ $\frac{21}{4}$ $\frac{21}{3}$

LEARN: Whatever number you multiply a denominator by, be sure to multiply the numerator by that same number to keep the new fraction equivalent to the fraction you started with.

Fractions and decimals

Look out for the signs often used for comparing fractions or other numbers:

$<$ This sign means 'is less than'.

$>$ This sign means 'is greater than'.

So, $\frac{1}{4} < \frac{1}{2}$ but $\frac{1}{2} > \frac{1}{4}$, whereas $\frac{2}{8} = \frac{1}{4}$.

HAVE A GO

1 Arrange these numbers in order, smallest first.

$1\frac{2}{3}$ $\quad \frac{7}{8}$ $\quad 1\frac{3}{4}$ $\quad \frac{9}{10}$ $\quad \frac{4}{9}$

2 Insert the correct sign $<$, $>$ or $=$ in the space between these fraction pairs.

a $\frac{7}{12}$ ________ $\frac{15}{24}$

b $\frac{6}{7}$ ________ $\frac{5}{8}$

c $\frac{48}{96}$ ________ $\frac{17}{34}$

Adding and subtracting fractions

If you have a calculation with more than one fraction in it, you need to make sure the denominators are the same, otherwise it will be difficult to work it out.

Look at these examples.

$1\frac{1}{2} + \frac{1}{4} = ?$

1. Change the mixed number to an **improper fraction**. $1\frac{1}{2} = \frac{3}{2}$
2. $\frac{3}{2}$ and $\frac{1}{4}$ have different denominators. 4 is a multiple of 2 and of 4.
3. Make the fractions have a denominator of 4. $\frac{3}{2} = \frac{3 \times 2}{2 \times 2} = \frac{6}{4}$

$1\frac{1}{2} + \frac{1}{4} = \frac{6}{4} + \frac{1}{4} = \frac{7}{4}$

4. Change the improper fraction to a mixed number. $7 \div 4 = 1$ remainder 3

$\frac{7}{4} = 1\frac{3}{4}$

$1\frac{1}{2} + \frac{1}{4} = 1\frac{3}{4}$

Do not forget that it can be easier to draw diagrams when working out fractions. $1\frac{1}{2} + \frac{1}{4} = 1\frac{3}{4}$ 

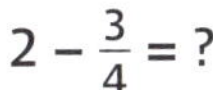

$2 - \frac{3}{4} = ?$

1. Change 2 into quarters. $2 = \frac{2}{1} = \frac{8}{4}$

2. $\frac{8}{4} - \frac{3}{4} = \frac{5}{4}$

3. Change the improper fraction to a mixed number. $5 \div 4 = 1$ remainder 1

$\frac{5}{4} = 1\frac{1}{4}$

$2 - \frac{3}{4} = 1\frac{1}{4}$ $2 - \frac{3}{4} = 1\frac{1}{4}$ 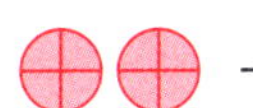 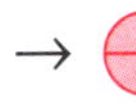

REMEMBER: Sometimes it helps to say the calculation in words: "eight quarters take away three quarters equals five quarters".

HAVE A GO

Calculate

1 $\frac{4}{11} + \frac{3}{11} =$ ____________

2 $2\frac{1}{4} - \frac{7}{8} =$ ____________

3 $\frac{3}{8} + \frac{3}{4} =$ ____________

4 $\frac{5}{6} - \frac{1}{5} =$ ____________

5 $3\frac{1}{2} + 5\frac{1}{4} =$ ____________

Fractions and decimals

Simplifying fractions

To understand what fractions mean, you usually write them as simply as possible. The fraction $\frac{42}{56}$ is difficult to understand or to draw. You can make it as simple as possible by reducing it to its **simplest** or **lowest terms**.

To do this, find a number that will divide exactly into both the numerator and the denominator (a common factor). Keep doing this until the **equivalent fraction** is as simple as possible, which means the numbers are as small as possible.

For example:

Reduce $\frac{42}{56}$ to its simplest terms.

42 and 56 have a common factor of 7. $42 \div 7 = 6, 56 \div 7 = 8$

$$\frac{42}{56} = \frac{6}{8}$$

6 and 8 have a common factor of 2. $6 \div 2 = 3, 8 \div 2 = 4$

$$\frac{6}{8} = \frac{3}{4}$$

$\frac{42}{56}$ is $\frac{3}{4}$ in simplest terms.

This is sometimes called **cancelling**.

$$\frac{\cancel{42}^{\cancel{6}^{3}}}{\cancel{56}_{\cancel{8}_{4}}} = \frac{3}{4}$$

First divide the top and bottom by 7 to make $\frac{6}{8}$. Then divide top and bottom by 2 to make $\frac{3}{4}$. When there are no more common factors to divide by, you have found the fraction in its simplest form.

HAVE A GO

Reduce these fractions to their simplest terms.

1 $\frac{5}{30}$ ______________________

2 $\frac{48}{96}$ ______________________

3 $\frac{70}{100}$ ______________________

4 $\frac{42}{49}$ ______________________

5 $\frac{32}{40}$ ______________________

Multiplying fractions

You can multiply fractions by multiplying the numerators together, multiplying the denominators together and then writing the answer in its simplest form like this:

$$\frac{2}{3} \times \frac{3}{8} \quad \begin{matrix}(2 \times 3 = 6)\\(3 \times 8 = 24)\end{matrix} \Longrightarrow \frac{6}{24} = \frac{1}{4}$$

By doing this, you have found two-thirds **of** three-eighths. Whenever you multiply fractions, you can say "**of**" instead of "times". This can help you work out some multiplying fractions calculations more easily.

Dividing fractions

Dividing fractions by whole numbers

What happens when you divide by 2? You find half or $\frac{1}{2}$ of something.

What happens when you divide by 4? You find a quarter or $\frac{1}{4}$ of something.

So ÷ 2 is the same as $\times \frac{1}{2}$ and ÷ 4 is the same as $\times \frac{1}{4}$.

This is true for any whole number, for example ÷ 498 is the same as $\times \frac{1}{498}$.

$$\frac{2}{3} \div 4 \Longrightarrow \frac{2}{3} \times \frac{1}{4} \Longrightarrow \begin{matrix}(2 \times 1 = 2)\\(3 \times 4 = 12)\end{matrix} \Longrightarrow \frac{2}{12} = \frac{1}{6}$$

LEARN: Any whole number can be written as an improper fraction. Just write the number as the numerator and 1 as the denominator.

Dividing fractions by fractions

The number 4 written as a fraction is $\frac{4}{1}$. So when you divide a fraction by 4 and you say it is the same as multiplying by $\frac{1}{4}$, you have turned the fraction $\frac{4}{1}$ upside down.

You can divide a fraction by any other fraction by turning the dividing fraction upside down and replacing the divide sign with a multiplication sign.

Another way to do this is to say "Keep it, change it, flip it", which means that you keep the first fraction, change the division sign to its inverse, so multiply, then flip the last fraction upside-down before multiplying.

See how this works:

$\frac{2}{3} \div \frac{3}{8} = \frac{2}{3} \times \frac{8}{3} \Longrightarrow \begin{matrix}(2 \times 8 = 16) \\ (3 \times 3 = 9)\end{matrix} \Longrightarrow \frac{16}{9} = 1\frac{7}{9}$

REMEMBER: Always make sure that your answer is written in its simplest form.

HAVE A GO

Calculate and simplify to the lowest terms.

1 $\frac{3}{5} \div \frac{1}{3} =$ ______________________

2 $\frac{5}{6} \div \frac{3}{8} =$ ______________________

3 $\frac{3}{4} \div 6 =$ ______________________

4 $\frac{7}{8} \div \frac{5}{12} =$ ______________________

5 $\frac{2}{7} \div 4 =$ ______________________

KEY FACTS

- $\frac{3}{4}$ ← numerator ← denominator
- The numerator and denominator in a common fraction are both whole numbers.
- To find a fraction of a number, divide the number by the denominator and then multiply the answer by the numerator.
- A mixed number is a mixture of a whole number and a fraction: $1\frac{3}{4}$.
- An improper fraction or 'top-heavy' fraction has a larger numerator than denominator: $\frac{9}{2}$.
- Equivalent fractions are equal to each other: $\frac{1}{3} = \frac{2}{6} = \frac{4}{12} = \frac{9}{27} = \frac{30}{90}$.
- Before completing a calculation with more than one fraction, make the denominators the same.
- To reduce a fraction to its simplest or lowest terms, divide the numerator and denominator by their highest common factor.
- To multiply fractions, multiply the numerators, multiply the denominators then simplify.
- To divide fractions, turn the second one upside down, then multiply and simplify.

EXAM TIP

In an exam, you may have to deal with fractions questions, so make sure you really know and understand the key facts. It can really help to draw simple diagrams of fractions. That is OK!

11 Decimal numbers

Decimal fractions

"What does 'decimal fractions' mean?"

When you write a decimal number, you have to put a dot to the right of the **digit** in the ones column. This dot is called a decimal point and shows the separation between whole numbers and numbers that are fractions between 0 and 1.

The digits to the left of the decimal point represent **whole numbers**: ones, tens, hundreds, thousands and so on.

The digits to the right of the decimal point represent parts of one whole: **tenths** ($\frac{1}{10}$s), **hundredths** ($\frac{1}{100}$s), **thousandths** ($\frac{1}{1000}$s) and so on.

The digits to the right of the decimal point are a decimal fraction.

An example of a **decimal fraction** is the number 0.256.

It is made up of $\frac{2}{10} + \frac{50}{100} + \frac{6}{1000}$ and is equal to $\frac{256}{1000}$.

REMEMBER: Decimal fractions always have a decimal point.

You can remind yourself about the value of digits in different columns by looking back at *Topic 1: Place value*.

Converting between decimal and common fractions

It is useful to know some decimal fractions and their common fraction equivalents. These are listed below.

$\frac{1}{10} = 0.1$	$\frac{4}{10} = \frac{2}{5} = 0.4$	$\frac{75}{100} = \frac{3}{4} = 0.75$
$\frac{2}{10} = \frac{1}{5} = 0.2$	$\frac{5}{10} = \frac{50}{100} = \frac{1}{2} = 0.5$	$\frac{8}{10} = \frac{4}{5} = 0.8$
$\frac{25}{100} = \frac{1}{4} = 0.25$	$\frac{6}{10} = \frac{3}{5} = 0.6$	$\frac{9}{10} = 0.9$
$\frac{3}{10} = 0.3$	$\frac{7}{10} = 0.7$	$\frac{10}{10} = 1.0$

If you need to change a **common fraction** like $\frac{3}{8}$ to a decimal fraction, it is very useful to remember that a common fraction can be thought of as a division calculation.

Look at this example:

Work out $\frac{3}{8}$ as a decimal fraction.

$\frac{3}{8}$ means **3 ÷ 8** or **3.0000 ÷ 8**

You can **estimate** that the answer will be less than **0.5** because $\frac{3}{8}$ is less than $\frac{4}{8}$, which is equivalent to $\frac{1}{2}$ or **0.5.**

$$\begin{array}{r}0.3\\8\overline{)3.0^{6}0000\ldots}\end{array}\qquad\begin{array}{r}0.3\,7\\8\overline{)3.0^{6}0^{4}000\ldots}\end{array}\qquad\begin{array}{r}0.3\,7\,5\\8\overline{)3.0^{6}0^{4}000\ldots}\end{array}$$

$\frac{3}{8}$ **= 0.375**

REMEMBER: Estimate your answer first, so that the decimal point does not end up in the wrong place.

LEARN: You can write as many zeros as you like at the end of a decimal number. 0.23 is the same as 0.230 000 000. 7.0 is the same as 7.000 000 0.

Sometimes you will have to change decimal fractions to common fractions or mixed numbers. It is important to think carefully about tenths, hundredths and thousandths.
See *Topic 1: Place value,* to remind yourself which column represents each of these.

The digits to the left of the decimal point (whole numbers) do not need to be changed to write a common fraction as a mixed number.

Digits to the right of the decimal point are the numerator in the fraction part of the mixed number.

If there is one digit to the right of the decimal point, then that digit represents the number of tenths:

If there are two digits to the right of the decimal point, then those digits represent the number of hundredths:

If there are three digits to the right of the decimal point, then those digits represent the number of thousandths:

$3.7 = \frac{37}{10}$

$3.75 = \frac{375}{100}$

$3.758 = \frac{3758}{1000}$

It is important to make sure your fraction is in its simplest terms.

$\frac{375}{100}$ can be simplified to become $3\frac{3}{4}$ or $8\frac{1}{4}$ as a mixed number.

See *Topic 10: Fractions* to remind yourself how to simplify fractions.

To arrange a group of decimal fractions in order, write them out one above the other, making sure the decimal points are all lined up above each other. Look at this example:

Write these numbers in order, smallest to largest.

3.241, 3.124, 3.412, 3.214, 3.421, 3.142.

The whole numbers are all the same.

Look at the tenths and hundredths digits coloured in red.

Put these in order of size: **12, 14, 21, 24, 41, 42.**

Now you can order the numbers easily.

The numbers in order are: 3.124, 3.142, 3.214, 3.241, 3.412, 3.421.

3.241
3.124
3.412
3.214
3.421
3.142

If you have numbers to order with a mixture of decimal fractions, such as **5.19, 5.1** and **5.278,** put zeros in the columns so that all numbeWrs can be compared easily: **5.190, 5.100, 5.278.**

Fractions and decimals

Add, subtract, multiply and divide decimal numbers

You will need to know how to add, subtract, multiply and divide decimal numbers in 11+ maths.

Decimals can be added and subtracted in exactly the same way as whole numbers.

Keep a clear idea of what each of the digits mean by reminding yourself of the place value columns (see Topic 1: Place value). It can be helpful to write the numbers you are adding or subtracting one underneath the other, always keeping the decimal point in the same column:

$$\begin{array}{r} 3.64 \\ +\ 5.29 \\ \hline 8.93 \\ \hline \end{array} \qquad \begin{array}{r} 7.65 \\ -\ 2.42 \\ \hline 5.23 \\ \hline \end{array}$$

To multiply decimal fractions, it can be useful to remember to say 'of' when you see a × sign.

For example:

In the calculation 0.2 × 0.2, you could ask yourself:

"What is two tenths of two tenths?"

First, find one tenth of 0.2 by dividing 0.2 by 10 (move the digits one column to the right) and then double it to find two tenths.

0.2 ÷ 10 = 0.02 0.02 × 2 = 0.04

Or you can just ignore the decimal points to begin with, multiply 2 × 2 = 4 and then say:

"There were two decimal places, so I have to put those back in my answer now":

0.2 × 0.2 = 0.04

To divide by a decimal fraction, it can be helpful to think of the decimal fraction as a common fraction.

For example:

In the calculation 5 ÷ 0.5, call 0.5 a half and say:

"How many halves in 5?"

There are two halves in one, so there are 5 × 2 = 10 halves in 5.

5 ÷ 0.5 = 10

For questions where the decimal fraction is not an obvious common fraction, then saying the question to yourself can still help.
For example, 6 ÷ 1.2 can be said as "How many 1.2s in 6?"

The question can then be solved using repeated addition:

1.2 + 1.2 + 1.2 + 1.2 + 1.2 = 6, therefore 6 ÷ 1.2 = 5

HAVE A GO

1 Change these common fractions into decimal fractions.

a $\frac{4}{5}$ ______ b $\frac{35}{100}$ ______

c $\frac{678}{1000}$ ______ d $\frac{7}{20}$ ______

e $\frac{1}{8}$ ______ f $\frac{3}{4}$ ______

2 Change these decimal numbers into mixed numbers.

a 5.8 ______ b 11.7 ______

c 23.006 ______ d 7.5 ______

e 4.12 ______

3 Arrange these decimal numbers in order of size.

5.123, 5.321, 5.14, 5.322, 5.032, 5.33, 5.3

______

4 Circle the correct answer to 0.2 × 0.2 × 0.2.

0.6 8 0.8 0.0008 0.0006 0.008

5 a Find the difference between 0.3 and 0.03. ______

b Divide 7 by 0.2. ______

EXAM TIPS

To help you remember how decimal numbers are organised in an exam, jot down the column headings between tens and thousandths.

When adding and subtracting decimal numbers, make sure the decimal points are above each other when you write the calculations down.

Always estimate the answer first by calculating with rounded numbers. Use this to check the decimal point is in the right place in your final answer.

KEY FACTS

- 0.37 is an example of a decimal fraction.
- In a decimal fraction, the digits to the right of the decimal point represent a fraction between 0 and 1. These can be shown as tenths, hundredths, thousandths and so on.
- A common fraction can be changed into a decimal fraction using division.
  $\frac{3}{8} \rightarrow 3 \div 8 = 0.375$
- A decimal number can be changed into a mixed number: $3.75 = \frac{375}{100} = 3\frac{3}{4}$
- Decimals can be used in addition, subtraction, multiplication and division calculations.

12 Percentages

LEARN: Percentage means 'out of 100'.

"I keep forgetting what the word 'percentage' means."

Per means 'out of' and cent means '100' (there are 100 **centimetres** in a metre; a century is 100 years; there are 100 cents in a dollar or euro).

These are some important **percentages** you should recognise:

$$100\% = \frac{100}{100} = 1 \text{ (one whole)}$$

$$75\% = \frac{75}{100} = \frac{3}{4} \text{ (three quarters)}$$

$$50\% = \frac{50}{100} = \frac{1}{2} \text{ (one half)}$$

$$25\% = \frac{25}{100} = \frac{1}{4} \text{ (one quarter)}$$

$$10\% = \frac{10}{100} = \frac{1}{10} \text{ (one tenth)}$$

You know how easy it is to divide by 10 (see *Topic 1: Place value*), so 10% is a particularly useful percentage for working things out.

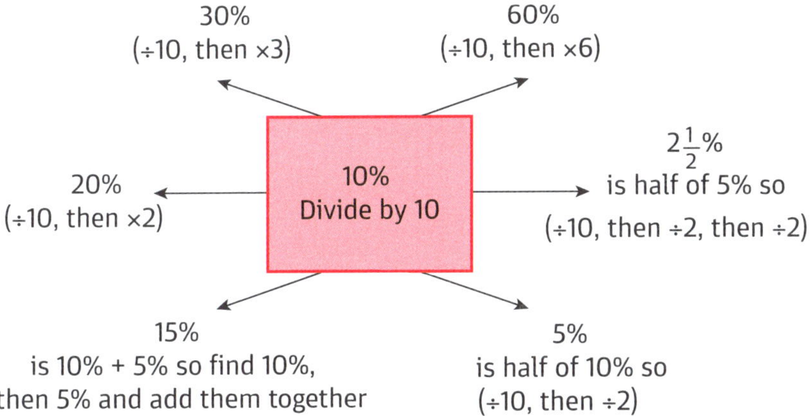

Fractions and decimals

It is also easy to find 1% by dividing an amount by 100. 1% means $\frac{1}{100}$. Once you know this, you can find all percentages this way: divide by 100, then multiply by the percentage amount you are trying to find.

Look at these examples:

Find 40% of 320.

1. Divide by 10 to find 10%. **320 ÷ 10 = 32**
2. Multiply by 4 to find 40%. **32 × 4 = 128**

40% of 320 = 128

Find 3% of £530.

1. Change £530 to pence. **53 000p**
2. Divide by 100 to find 1%. **53 000 ÷ 100 = 530p**
3. Multiply by 3 to find 3%. **530 × 3 = 1500 + 90 = 1590p**
4. Change back to pounds. **£15.90**

To increase a given amount by a percentage, you find the given percentage and add it on to the original amount.

To decrease a given amount by a percentage, you find the given percentage and subtract it from the original amount.

For example:

What is the price of a £25 pair of jeans with 20% off in the sale?

1 Find 10% of the price. £25.00 ÷ 10 = £2.50

2 20% is 2 × 10% so double the amount. £2.50 × 2 = £5.00

3 Take £5 off the original price. £25.00 – £5.00 = £20

The new price is £20.

HAVE A GO

1 Find 4% of £420. ______

2 What is 0.45 as a percentage? ______

3 20% of children in a school of 420 pupils were absent.
How many were present? ______

4 In a sale, everything is reduced by 15%.
What is the cost of a jacket which normally costs £35? ______

5 The cost of a car is increased by 16% at the beginning of August.
In July, the car cost £13 000. What does it cost in August? ______

KEY FACTS

- Percentages are all based on something being divided into 100 equal parts. 100% of something means all of it; 50% means one half of it; and 25% means one quarter of it.
- To find 1% of an amount, divide it by 100.
- For 11+ maths, you can work out most percentages by finding 10% first.

13 Ratio and proportion

Ratio

A **ratio** shows how a total amount can be shared out in unequal parts. A colon between two or more numbers is used to show ratio in maths, for example 3 : 2.

The picture shows red and white beads in the ratio 3 : 2. This can be said as 'three red beads to two white beads' or '3 to every 2'.

REMEMBER: To describe a ratio say 'to' or 'to every'.

When you share something in a given ratio, you must first add up the numbers in the ratio to find out how many equal parts you need to work with. This is like finding out the denominator in a fraction.

REMEMBER: A ratio is used to compare two or more numbers or quantities.

You can also write ratios with more than two numbers. 4 : 5 : 6 might describe a recipe where you need 4 mushrooms for every 5 tomatoes and 6 eggs.

Or 4 : 5 : 6 might be a way of sharing £15 between three sisters. There are 15 parts altogether (4 + 5 + 6 = 15) so £15 would be shared in the ratio £4 : £5 : £6. If the sisters were sharing £45 in the same ratio, they would each get three times as much, because £45 is three times as much as £15 – they would get £12 : £15 : £18.

LEARN: To describe a proportion it can be helpful to say 'in every' or 'out of every'.

Proportion

If the ratio of 3 red beads to every 2 white beads is used to arrange 5 beads in a pattern, then the **proportion** of red beads is 3 in every 5, or 3 out of every 5 and the proportion of white beads is 2 in every 5, or 2 out of every 5:

LEARN: When you share out an amount using a given ratio, then each share is a proportion of the total amount.

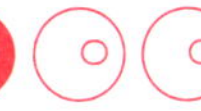

$\frac{3}{5}$ of the beads are red. $\frac{2}{5}$ of the beads are white.

Look at these examples:

Fractions and decimals

£4.80 is shared between Adam, Aisha and Bethany in the ratio 3:1:2. What proportion of the money does Bethany receive?

1. Find the total number of equal parts.

   3 + 1 + 2 = 6

2. Bethany is third in the list of names so she gets the third portion of the ratio.

   She gets 2 out of 6 parts.

   2 out of 6 = $\frac{2}{6} = \frac{1}{3}$

Bethany gets $\frac{1}{3}$ **of the money.**

£4.80 is shared between Adam, Aisha and Bethany in the ratio 3:1:2. What amount of money does Bethany receive?

1. Find the total number of equal parts.

   3 + 1 + 2 = 6

2. Divide the amount to be shared by the number of equal parts.

   £4.80 ÷ 6 = £0.80

   Each part is worth £0.80.

3. Bethany is third in the list of names so she gets the third number of parts.

   She gets 2 parts.

   £0.80 × 2 = £1.60

Bethany receives £1.60.

Scale factor

If you imagine drawing a picture of your hand, you would use a whole piece of A4 paper to draw around your hand, fingers and thumb. If you wanted to draw your whole body, you would be joining lots of pieces of paper together. It would not be very practical! Instead, you can represent your body by using a **scale factor**. You might use 1 cm on the paper to represent 10 cm of your body. You would write this scale factor like this: 1 : 10, which is a ratio.

Every 1 cm on your drawing represents or stands for 10 cm in real life.

When you reduce an accurate drawing of a larger thing so that it fits onto a piece of paper, you use a scale factor ratio to do this. Scale factors are used for recording distances on maps and in atlases, for making accurate models and for plans when designing buildings.

Look at this example:

My map has a scale of 1 : 150 000. How far is it from my house to the airport if it measures 6 cm on the map?

1 cm on the map represents 150 000 cm on the ground.

1 cm on the map represents 1500 m on the ground (divide by 100 to turn **centimetres** into **metres**).

1 cm on the map represents 1.5 km on the ground (divide by 1000 to turn metres into **kilometres**).

6 cm on the map represents **6 × 1.5 km = 9 km.**

It is 9 km from the house to the airport.

LEARN: You could just divide by 100 000 to go straight from centimetres to kilometres, but going step-by-step can make it easier.

Fractions and decimals

HAVE A GO

1 Harvey watched 45 cars go past his house. 3 in every 5 cars were silver. How many silver cars did he see? ______

2 £6.40 is shared between Rose, Sophie, Flynn and Ahmed in the ratio 1:2:5:8.

a How much money does Flynn receive? ______

b What proportion does Sophie receive? ______

3 A map is drawn to a scale of 1:2000. What distance in metres would be represented by 5 cm on the map? ______

4 In a youth group, there are 91 children. There are 4 boys to every 3 girls. How many girls are there? ______

EXAM TIP

If you find 11+ practice questions about ratio, proportion and scale factor difficult to understand, be sure to go over Topic 10: Fractions and Topic 25: Metric and imperial units of measurement first. Try to imagine what the problems are about and work in careful steps. These questions often have more marks because there is a lot of working out to do.

KEY FACTS

- A ratio is used to compare two numbers or quantities.
- Ratio is said as 'to every' or 'for every' and is written using a colon between the parts, for example 2:1. This is the ratio '2 to 1'.
- When you share out an amount using a given ratio, then each share is a proportion of the total amount.
- Proportion is a fraction of the whole amount.
- You might be asked to find the proportion (the fraction or percentage) of an amount shared, or the amount someone gets when something is shared.
- A scale factor is a ratio between a real-life measurement and a scaled representation.

Handling data

14 *Organising and comparing information*

Data

Data can be numbers or **values** or words and is usually collected by observing, questioning or measuring. In 11+ maths, you need to be able to understand different ways of showing data, such as line **graphs**, block graphs, **pie charts**, tables of information and **Venn diagrams**.

LEARN: Data is the word that describes collections of information.

You also need to be able to answer questions about data shown in these ways. If you have to draw graphs or charts yourself, you must be very careful to choose a sensible way of showing the numbers from your data. Organising your information into tables may show patterns or similarities in your data.

REMEMBER: A graph is a type of picture that shows data.

Graphs

Graphs can have bars, lines or pictures representing the data.

Every graph has two lines, the ***x*-axis** and ***y*-axis** (plural: **axes**), which join or intersect at the point (0,0) called the **origin**, like this:

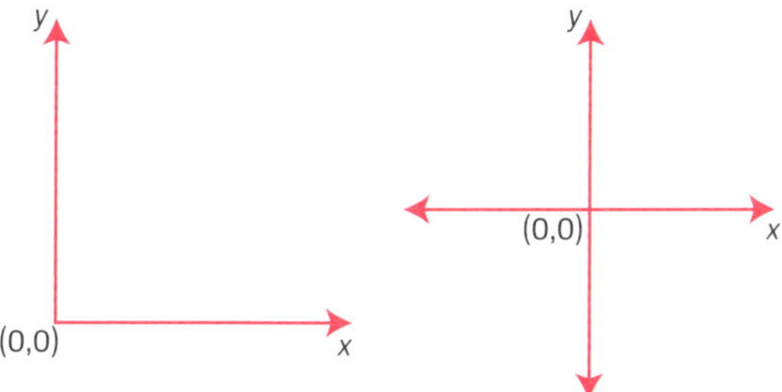

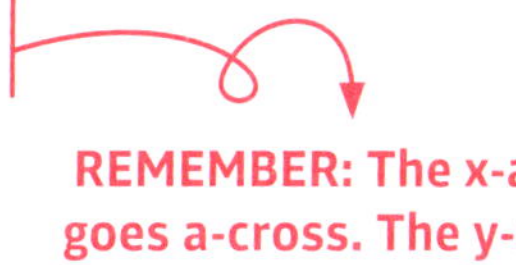

REMEMBER: The x-axis goes a-cross. The y-axis goes up in the sky.

x and *y* are mystery amounts in maths and you must label the axes to make it clear what they represent.

The *x*-axis is always **horizontal** (straight across, like the horizon).

The *y*-axis is always **vertical** (straight up and down).

Look at the bar graph. It shows the **percentage** of children in a school who own different pets.

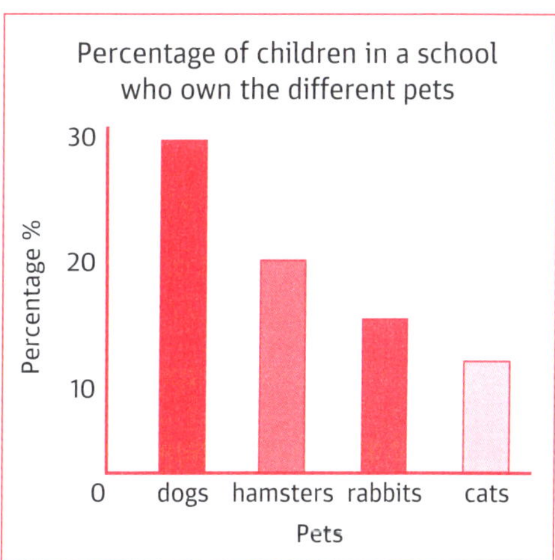

Handling data

Pie charts

A pie chart is a circle divided into sections to show how something is shared or divided into groups. Look at this pie chart:

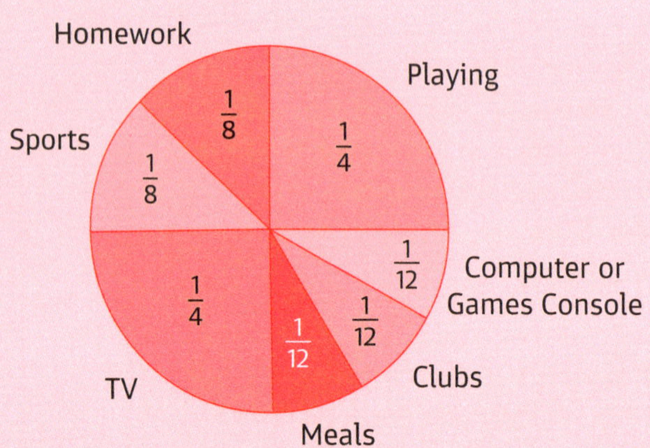

It shows how much time children in Year 6 spend doing different activities between 4 **p.m.** and 8 **p.m.**

From the chart, you can see, for example, that they spend $\frac{1}{8}$ of their time doing homework, $\frac{1}{4}$ of their time playing and $\frac{1}{12}$ of their time going to clubs.

The following facts can help you answer many questions about pie charts.

See *Topic 17: 2D shapes: circles, angles and bearings* for more information about circles and **angles**.

LEARN: There are 360° in a circle, 180° in a **semicircle** and 90° in a **right angle**.

- 360° represents 100% = 1
- 180° represents 50% = $\frac{1}{2}$
- 90° represents 25% = $\frac{1}{4}$
- 270° represents 75% = $\frac{3}{4}$
- 45° represents $12\frac{1}{2}$% = $\frac{1}{8}$

Look again at the pie chart on page 56. What percentage of their time do the children spend watching TV? What about doing sports?

A quarter of the pie chart represents TV, so they watch TV for 25% of the time.

An eighth of the pie chart represents sports, so they do sport for $12\frac{1}{2}$ % of the time.

Venn diagrams

Venn diagrams are used to organise data in rings. Each ring is labelled to show which items of data can be included in it and which cannot. Where two or more rings overlap, the data that can be included in both or all rings is written in the overlapped section.

The table shows the type of pets that 25 children in Year 6 have:

Pet	Number of children
Dogs	6
Cats	12
Other pets	8
No pets	6
Total	**32**

REMEMBER: Venn diagrams are diagrams in which the data is organised into rings.

This data is shown in a Venn diagram on the next page and gives us some more information.

HAVE A GO

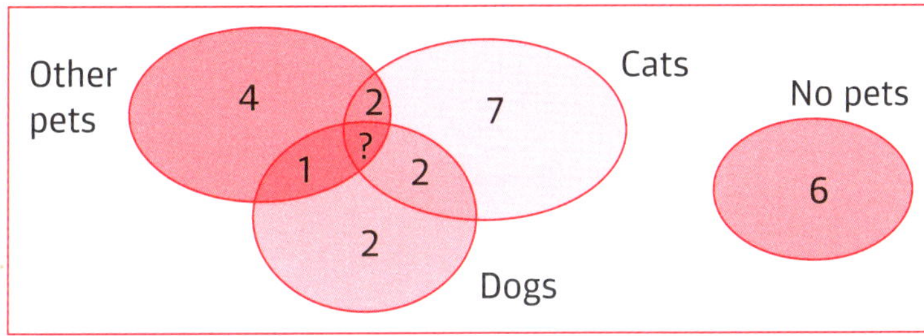

1 Look at the Venn diagram above. How can you find what number should go in the overlapping part of the three rings so that the information this shows is correct?

2 Class 6 surveyed 80 children, asking them their favourite food. Here is the pie chart of their results. How many children liked fish and chips best?

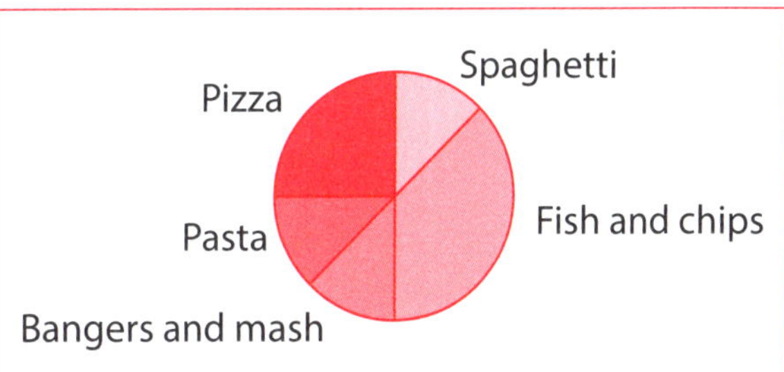

____

HAVE A GO

3 Maryam went cycling and timed herself. Later, she made this graph.

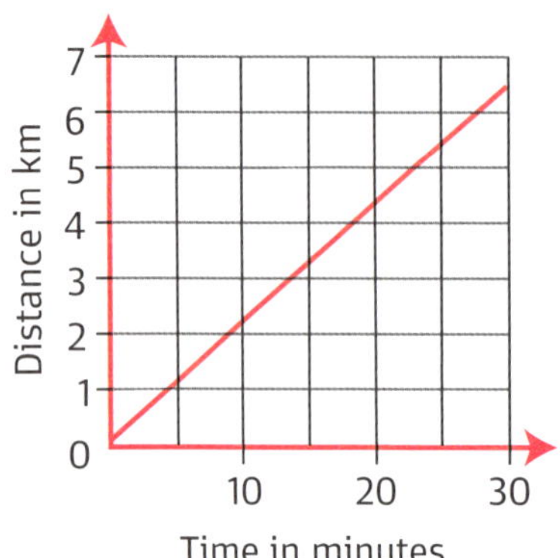

What was her speed in km/h (kilometres per hour)?

____

4 34 children take part in an after-school club. This Venn diagram shows how many of them belong to the art group (A) and the computer group (C). How many children do not belong to the computer group?

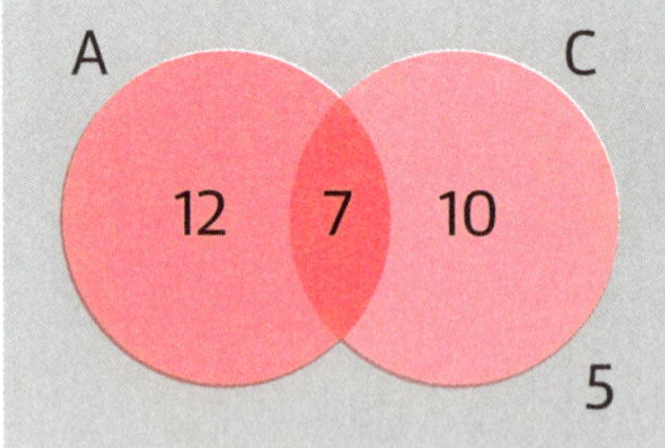

____

KEY FACTS

- Data describes collections of information.
- Graphs have an *x*-axis (horizontal) and a *y*-axis (vertical).
- Pie charts are circles split up into sections to represent different groups of data.
- Venn diagrams use rings to organise data, but only data meeting the given rule for a ring can be included in it.

EXAM TIP

In your exam, make sure you read charts, tables and their keys carefully so that you know what information you are trying to find. Finding the totals and jotting them down first can make answering the questions quicker, rather than starting again each time.

15 Mean, median, mode and range

Mean, median and mode

There are three different types of **average** of a set of **data**: the **mean**, the **median** and the **mode**.

> **LEARN:** An average is a single number or value that represents a set of data.

The mean is found by adding together all the numbers (or **values**) in the set of data and dividing by how many numbers there are in the set.

The median is the middle number (or value) when the numbers in the set are put in order of size. If there are two middle numbers, you find the mean of those two numbers by adding them up and dividing by two. This is the median.

The mode is the number (or value) in the data that comes up most often.

Range

The **range** of a set of numbers is the difference between the smallest number and the largest number.

Look at this example:

Find the range, mode, median and mean for this set of test results.

7	10	7	5	9	9	10	6	9

Range: Highest score is 10, lowest score is 5. 10 – 5 = 5
The range is 5.

Mode: 9 appears more times than any other score.
The mode is 9.

Median: You need to arrange the scores in order.

5	6	7	7	9	9	9	10	10

The middle number is 9 so the median is 9.

Mean: 5 + 6 + 7 + 7 + 9 + 9 + 9 + 10 + 10 = 72

There are nine scores with a total of 72. 72 ÷ 9 = 8
The mean is 8.

HAVE A GO

Find the range, mode, median and mean for this set of shoe sizes in a Y6 class.

4, 5, 6, 6, 4, 4, 6, 5, 8, 4, 3, 4, 4, 7, 5, 5, 3, 5, 6, 6

Range: ______________ Mode: ______________

Median: ______________ Mean: ______________

KEY FACTS

- Average – a single number or value that best represents a set of data.
- Mode – the value that appears the most often.
- Median – the value in the middle when put in size order.
- Mean – the result of dividing the total of all the values by the number of values. It is often just called the average of a set of numbers.
- Range – the difference between the largest and the smallest value.

16 Probability

The **probability** of something happening is the likelihood or chance of it happening.

If something is certain to happen, its probability is 1.

If something is impossible, its probability is 0.

If an event is neither certain nor impossible, its probability is somewhere between 0 and 1 and is often written as a **fraction**.

For example, imagine tossing a coin and it landing heads up. There is a 1 in 2 chance of this happening, so its probability is $\frac{1}{2}$:

REMEMBER: Probability means 'chance' or 'possibility'.

The number of correct outcomes ⟶ 1 ⟵ Heads: only one correct outcome

The number of possible outcomes ⟶ 2 ⟵ Heads or tails: two possible outcomes

Sometimes, as with throwing two dice, there may be many different combinations that are possible.

Each die (or dice) has six numbers, so if you throw two dice you have 6 × 6 possibilities: 36 in all. There is only one way to score a total of 2 with two dice (1 and 1), so the probability of scoring 2 with two dice is 1 in 36 or $\frac{1}{36}$.

Look at this example:

If you roll two dice, what is the probability that the total score is 6?

Here are the ways you could score 6:

Dice 1	1	5	2	4	3
Dice 2	5	1	4	2	3

The total 6 can be made in 5 different ways.

There are 36 scores, so the probability of the total being 6 is $\frac{5}{36}$.

In 11+ maths, you may have to find the probability of picking balls or numbers out of a bag, throwing a die or dice, tossing coins or choosing playing cards from a pack.

It is possible to have weighted or unfair dice and coins, so very often a question will tell you that the dice or coins are fair. It means all the possible outcomes are equally likely.

LEARN:

A pack of playing cards has four suits: hearts, diamonds, clubs and spades.

Each suit has 13 cards: ace (1), 2, 3, 4, 5, 6, 7, 8, 9, 10, jack, queen and king.

Altogether there are 52 cards. (Do not count the jokers!)

HAVE A GO

1 If a coin is thrown 10 times, how many times is it likely to land on heads? ______

2 In a bag there are six balls numbered 1 to 6.
What is the probability that you would take out:

a the ball numbered 5? ______

b a ball that has an even number on it ______

3 If you roll two dice, what is the probability that the total of the two dice is either 3 or 4? ______

4 In a bag there are 3 red balls and 4 blue balls. Circle the probability of picking:

a a red ball: $\frac{3}{8}$ $\frac{3}{7}$ $\frac{3}{6}$ $\frac{3}{5}$ $\frac{3}{4}$

b a blue ball: 0 $\frac{1}{2}$ $\frac{4}{7}$ $\frac{3}{7}$ $\frac{3}{4}$

c a yellow ball: 0 $\frac{1}{2}$ $\frac{4}{7}$ $\frac{3}{7}$ $\frac{3}{4}$

5 What is the probability of picking a royal card (jack, queen, king) from a pack of cards? ______

EXAM TIP

It can be helpful to write down all the possibilities before working out a probability but be methodical so that you are sure to get all the combinations!

KEY FACTS

- The probability of something happening is the likelihood or chance of it happening.
- A probability of 1 means an event is certain to happen.
- A probability of 0 means an event is impossible or certain not to happen.
- All other probabilities are given as a fraction between 0 and 1.

Shape and space

17 2D shapes: circles, angles and bearings

Circles

You need to know the names of certain parts of a circle.

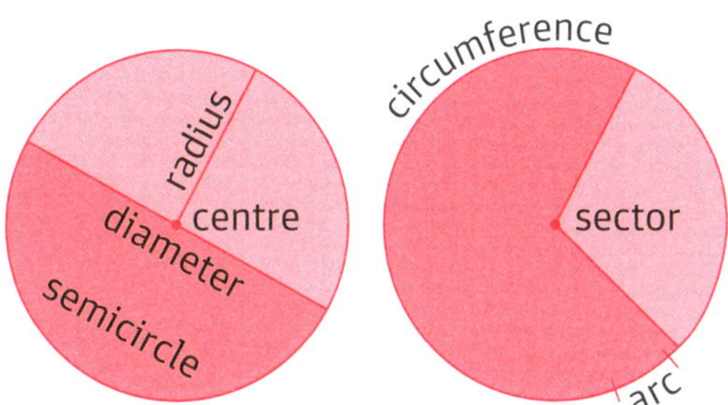

LEARN: If you draw a circle using a pair of compasses, the hole left by the sharp point is the centre or the middle of the circle.

Concentric circles share the same centre point but the **radius** of each circle is different.

It is important to know the relationship between the **diameter** and radius of a circle:

the diameter is two times the radius $d = 2 \times r$

the radius is half the diameter $r = \frac{1}{2}$ of d or $\frac{d}{2}$

A circle has an infinite (never-ending) number of radii (or radiuses) and an infinite number of diameters, which means that you can draw a radius or a diameter anywhere on a circle as long as it goes to the **circumference** and from or through the centre of the circle.

An **arc** is part of the circumference. A **sector** is part of a circle, between two radii.

Angles

An **angle** tells us how far something turns or rotates.

Angles are measured in **degrees**. There are 360 degrees (360°) in a full turn.

When a circle is divided into quarters, four **right angles** (90°) are made. Each right angle is the same as a quarter turn.

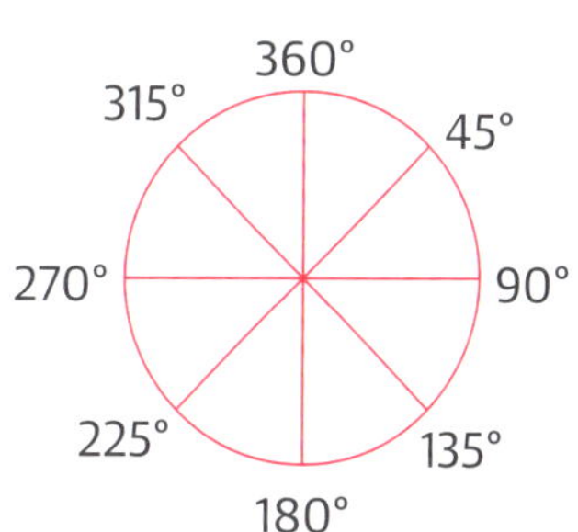

Shape and space

It is very important that you recognise a right angle. It looks like the corner of a **square** or a piece of paper.

An angle of 180° is a straight line. It can be thought of as two right angles (90°) together. Your **protractor** (a tool for measuring angles) will probably have 180° as its largest amount. It is half a complete turn or the angle in a **semicircle**.

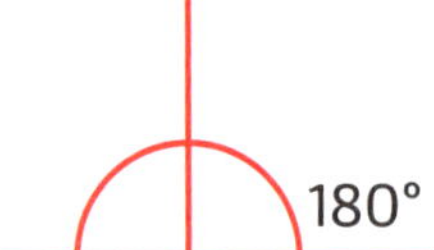

Angles less than 90° are called acute. Acute means sharp.

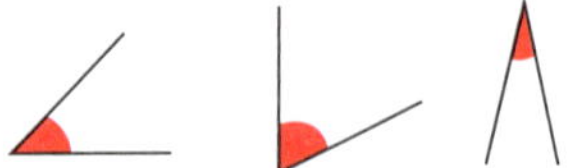

Angles greater than 90° but less than 180° are called obtuse. Obtuse means blunt.

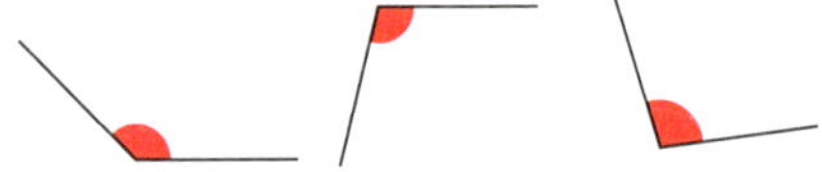

Angles greater than 180° are called reflex.

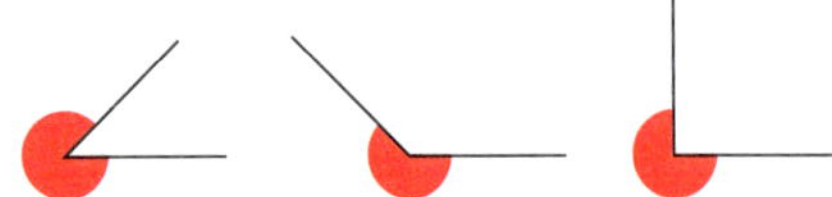

Every right angle, **acute angle** or **obtuse angle** has a matching or **complementary reflex angle**. The two angles add up to 360°, a complete turn.

To measure angles accurately, you need to be able to use a protractor, a maths tool consisting of a transparent semicircle with two scales, one **clockwise** and one **anticlockwise**, each with the degrees between 0 and 180 marked on it *(see Topic 26: Reading scales)*.

Two lines are **perpendicular** if they are at right angles to each other. One line is perpendicular to the other.

Line AC is at right angles to line BC.

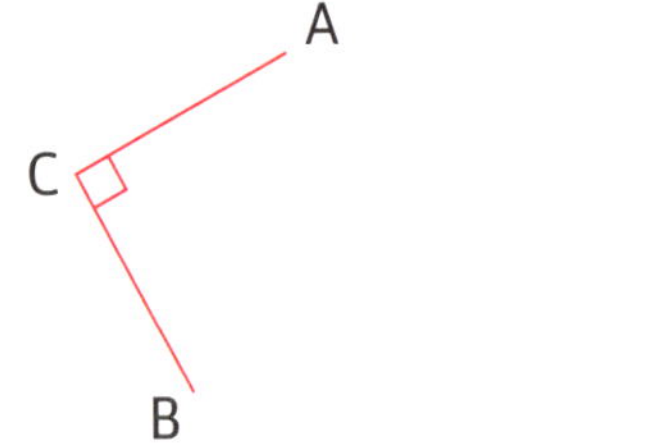

Line AC is perpendicular to line BD.

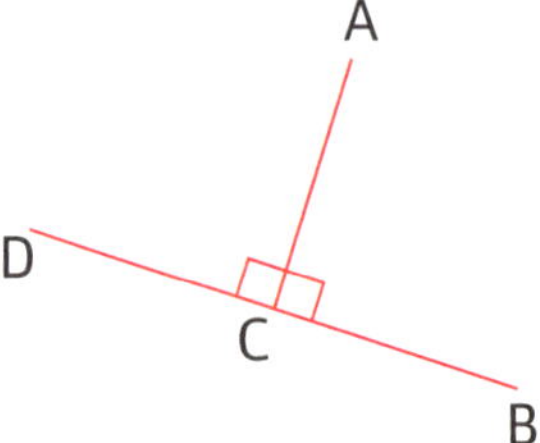

Notice how lines are named with a letter at the start and a letter at the end.

Two lines are **parallel** if they travel in the same direction and are the same distance apart all along their lengths.

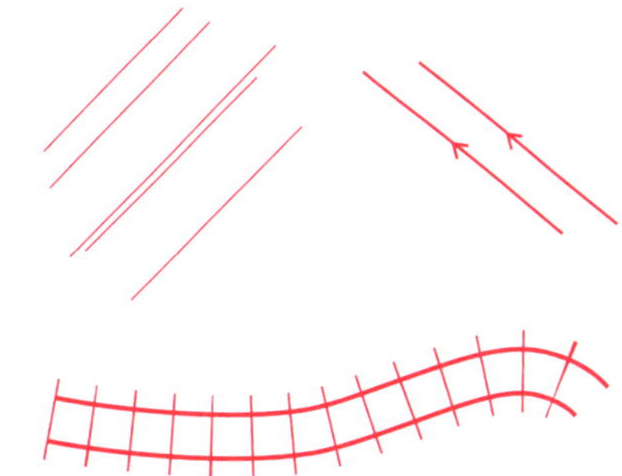

Parallel lines never meet and therefore will never make an angle between them.

Parallel lines do not have to be the same length.

Parallel lines do not have to be straight.
Think of railway tracks!

In diagrams, you may see lines with arrow symbols on them. This means that they are parallel.

Bearings

The diagram on the right shows the main directions marked on a compass.

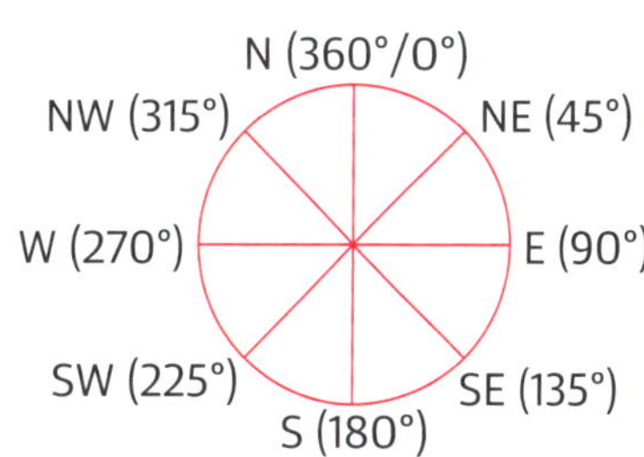

A **bearing** is the angle between the direction north and the direction in which something is travelling. You always measure bearings clockwise from north.

See how the compass points match the degrees shown on the circle on page 63. These are the bearing measurements for the eight main compass points.

HAVE A GO

1 The radius of a circle is 1.9 cm. What is its diameter? ________________

2 If you are facing SE and you turn 135° anticlockwise, in which direction are you now facing? ________________

3 In a right angle, one angle of 37° is cut off. What does the remaining angle measure? ________________

4 Jack draws a circle and measures a turn of 100°. What is the complementary angle to this? ________________

5 What is the smaller angle between the numbers 4 and 8 on a clock face? ________

EXAM TIP

There are a lot of new words to learn in this section. Make some flashcards well before the exam with each word you are not sure about on one side and the explanation on the other to help you practise these. In the exam, it may help you to visualise questions about circles, angles and bearings if you draw sketches of what is going on.

KEY FACTS

- There are 360° in a full turn.
- An acute angle is less than 90°.
- A right angle is 90°.
- An obtuse angle is between 90° and 180°.
- A reflex angle is greater than 180°.
- Every acute, right or obtuse angle has a **complementary** reflex angle to make a complete turn.
- Perpendicular lines are at right angles to each other.
- Parallel lines never meet but run the same distance apart from each other for their entire length.
- A compass is used to find a bearing: the angle between north and the direction in which something is travelling.
- The eight main points on a compass are N, NE, E, SE, S, SW, W and NW.

18 2D shapes: triangles

Types of triangle

"Triangles all look very similar to me."

Triangles can be lots of different shapes, but they all have three sides and three angles.

They can be named according to their **sides** and sorted into different groups.

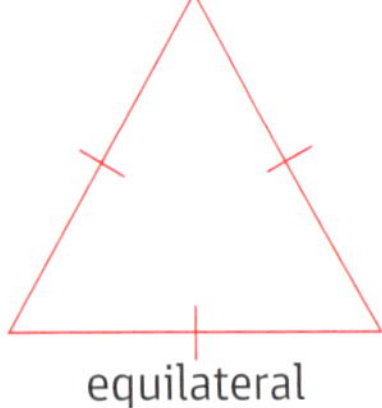

An **equilateral triangle** has three equal sides and three equal angles.

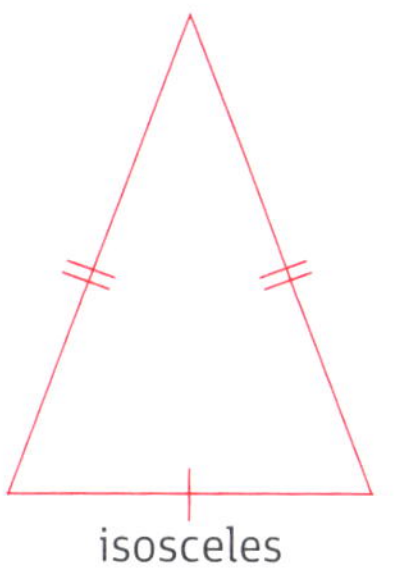

An **isosceles triangle** has two equal sides and two equal angles.

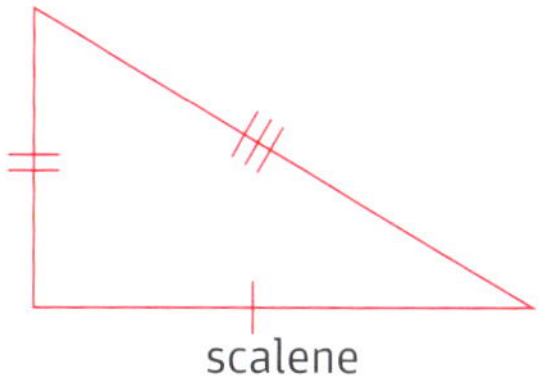

A **scalene triangle** has no equal sides and no equal angles.

Notice how the sides can be labelled with short lines or 'whiskers' to show how they compare with each other.

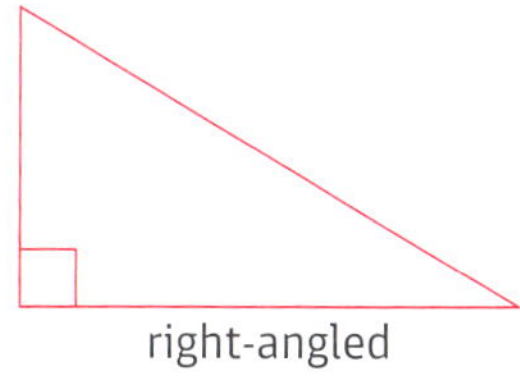

Triangles can also be named according to their angles.

A **right-angled triangle** has one right angle.

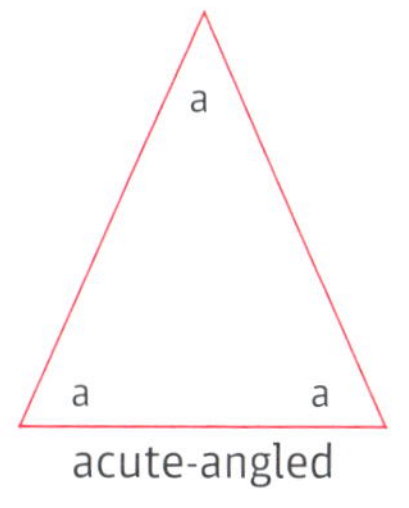

An **acute-angled triangle** has three acute angles.

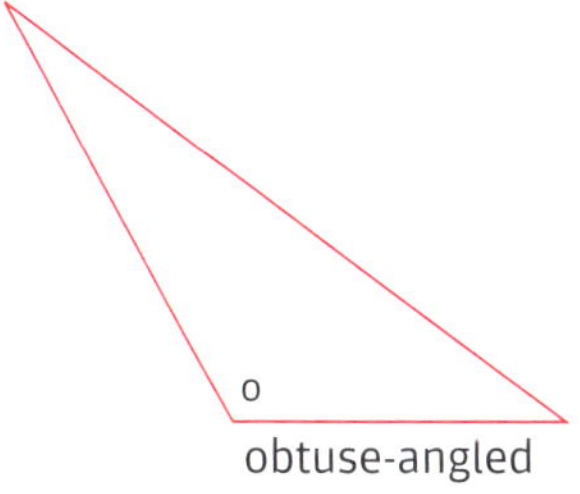

An **obtuse-angled triangle** has one obtuse angle.

If you tear off the three angles or corners of any paper triangle and place the angles together along a straight line or **edge**, they will fit together to make a straight angle or 180°.

LEARN: The three angles of any triangle always add up to 180°.

Area of a triangle

To find the **area** of any triangle, imagine it in a **rectangle**. It fills half the space of the rectangle. Find the area of the rectangle (length of its base × the length of its height) and then halve that amount.

Area = $\frac{1}{2}$ × (base × height) or A = $\frac{1}{2}$ (b × h)

The area of a triangle (*A*) is its base (*b*) multiplied by its height (*h*), all divided by 2.

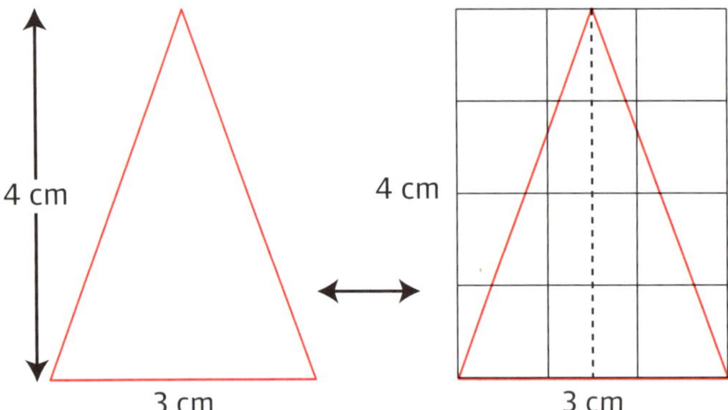

The area of this triangle is:

$\frac{1}{2} \times (3\,\text{cm} \times 4\,\text{cm}) = 6\,\text{cm}^2$

See *Topic 20: Perimeter and area for a description of the units of measurement used to represent area*.

HAVE A GO

1 Two angles of a triangle measure 80° and 45°. Find what the third angle measures. ____________________

2 Complete the sentence. Circle the correct letter.

An equilateral triangle is a kind of ____________________

A right-angled triangle **B** acute-angled triangle **C** obtuse-angled triangle

3 What is the name of a triangle with three different sides and three different angles? ______

4 A triangle's height measures 7 cm and its base measures 12 cm. What is its area? ______

KEY FACTS

- Equilateral triangle – three equal sides and angles.
- Isosceles triangle – two equal sides and angles.
- Scalene triangle – no equal sides or angles.
- Right-angled triangle – one right angle.
- Acute-angled triangle – three acute angles.
- Obtuse-angled triangle – one obtuse angle.
- Area of a triangle = $\frac{1}{2}$(base × height).
- The three angles of any triangle always add up to 180°.

19 2D shapes: quadrilaterals and other polygons

REMEMBER: The word quadrilateral can be used for all four-sided 2D shapes.

Quadrilaterals

Quadrilaterals have four sides and also have four angles.

If you join up the opposite angles of a quadrilateral, you form its **diagonals**.

The quadrilaterals you need to know are on the next page.

"The names of some of these four-sided shapes are hard to remember."

This is where a **mnemonic** may help. Try making up a rhyme or silly story to help you learn the names. Practise drawing and labelling the different types of quadrilaterals until you know them.

If you draw and cut out different paper quadrilaterals and for each one, tear off its four angles and put them together, they will always make a complete angle of 360°.

LEARN: The four angles of any quadrilateral always add up to 360°.

You need to know the names and properties (what makes them special) of these quadrilaterals:

- A **square** has four equal sides, four right angles and four lines of symmetry.

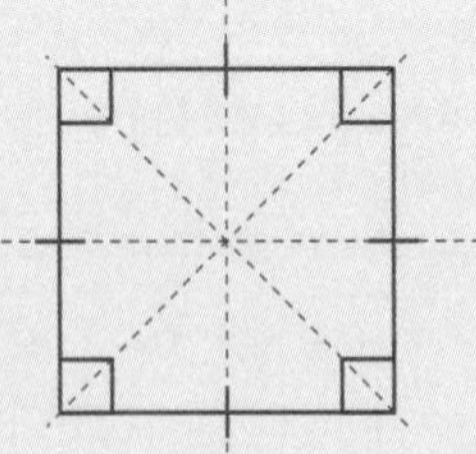

- A **rhombus** also has four equal sides, but two equal acute angles, two equal obtuse angles and two lines of symmetry.

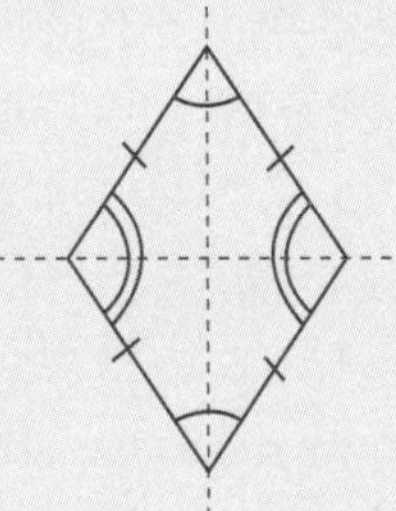

- A **rectangle** has two pairs of **parallel** sides of equal length, four right angles and two lines of symmetry. Rectangles are sometimes called oblongs.

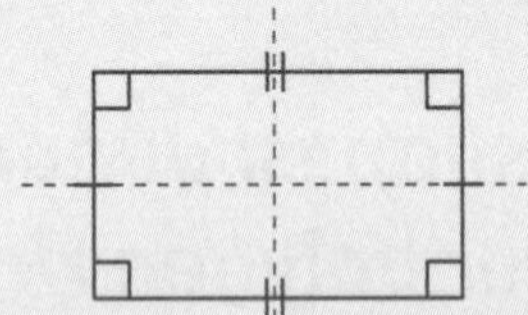

- A **parallelogram** has two pairs of parallel sides, two equal acute angles and two equal obtuse angles. It has no lines of symmetry.

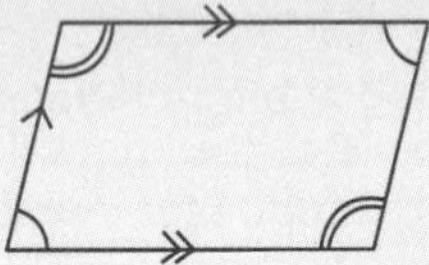

- A **kite** has two pairs of sides that are **adjacent**, or next to each other. It has one pair of opposite equal angles and one line of symmetry.

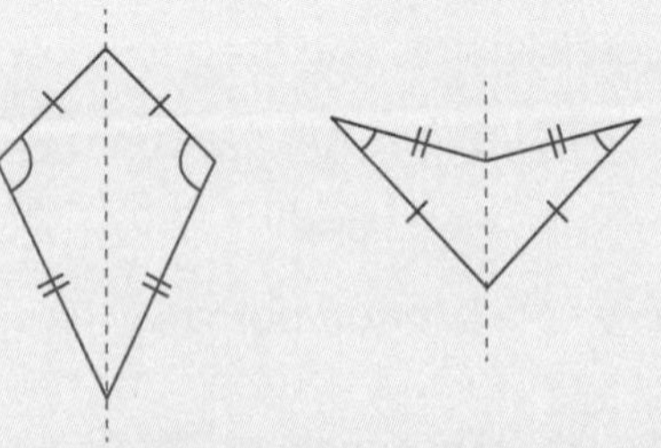

- A **trapezium** looks like a triangle with its head chopped off! Two of its sides are parallel. It has no lines of symmetry.

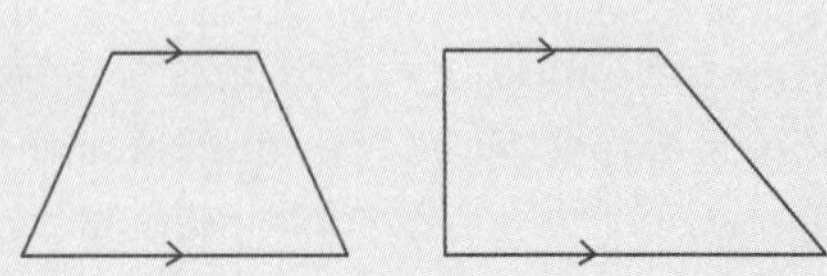

Polygons

The sides of a **regular polygon** are the same length and all angles are equal. An **irregular polygon** has sides and angles that are not the same.

You need to know the names of these regular polygons and be able to recognise them:

LEARN: A polygon is any 2D shape with three or more straight sides.

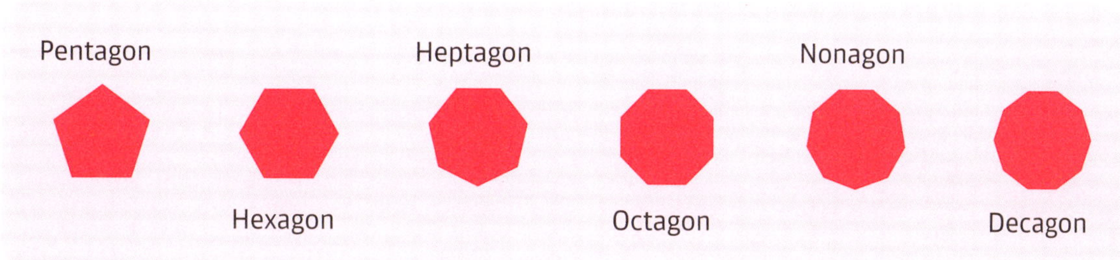

A **pentagon** has five sides.

A **hexagon** has six sides. The **cross-section** of some pencils is hexagonal.

A **heptagon** has seven sides. A 20p or 50p coin is a heptagon.

An **octagon** has eight sides. An octopus has eight tentacles and in music, an octave is eight notes in a scale.

A **nonagon** has nine sides.

A **decagon** has ten sides.

REMEMBER: The more sides a polygon has, the more it resembles a circle.

You may need to calculate the interior (inside) angles of a regular polygon. To do this, imagine a **polygon** you are thinking about divided into triangles. Each triangle's angles add up to 180°. For each polygon, count the sides, subtract 2 and multiply by 180. This will give you the total of the interior angles. Now all you have to do is divide this total by the number of the polygon's sides to give you one angle size.

Find the interior angle of a regular octagon.

1. An octagon has 8 sides. Subtract 2.	$8 - 2 = 6$
2. Multiply 6 by 180.	$6 \times 180 = 1080$
3. Divide 1080 by 8.	$1080 \div 8 = 145$

Each interior angle of a regular octagon measures 145°.

HAVE A GO

1 A quadrilateral has angles of 110°, 120°, 80° and one more. What does it measure in degrees? ____________

2 What is one of the interior angles of a regular pentagon in degrees? ____________

3 How many lines of symmetry does a regular hexagon have? ____________

4 Find in degrees the size of one of the angles marked z. ____________

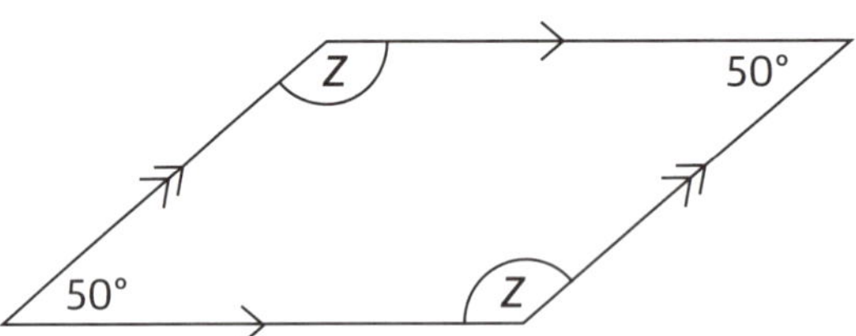

5 Which two shapes are used to make this tile pattern?

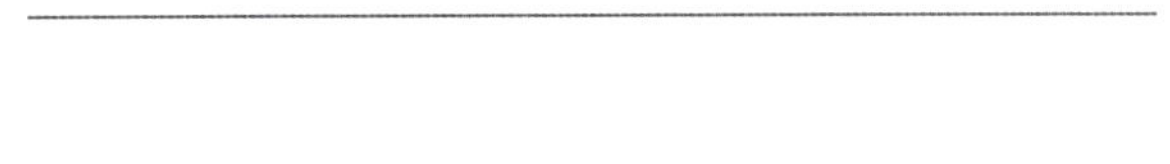

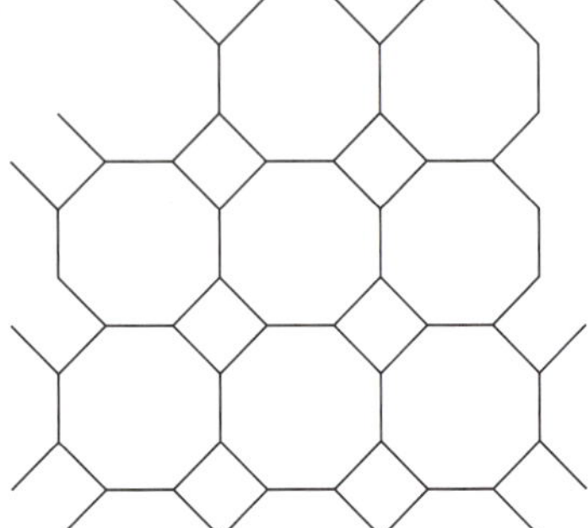

KEY FACTS

- Quadrilaterals are four-sided 2D shapes.
- Polygons are 2D shapes with three or more straight sides.
- The sides of a regular polygon are the same length and all angles are equal. All other polygons are irregular.
- Polygons are named according to the number of sides.

Name of polygon	Number of sides
triangle	3
quadrilateral	4
pentagon	5
hexagon	6
heptagon	7
octagon	8
nonagon	9
decagon	10

Shape and space

20 Perimeter and area

Perimeter

"I get perimeter and area muddled."

Think of a hedge going round the **edge** of a really muddy field. The hedge makes the perimeter and is 'so many' **metres** long. The field it encloses has an area of 'so many' metres square.

LEARN: Perimeter is a length. It might be measured in millimetres (mm), centimetres (cm), metres (m) or kilometres (km).

The perimeter of a 2D shape is the total distance round the edge of the shape.

To find the perimeter of this rectangle, add all the measurements of the sides together.

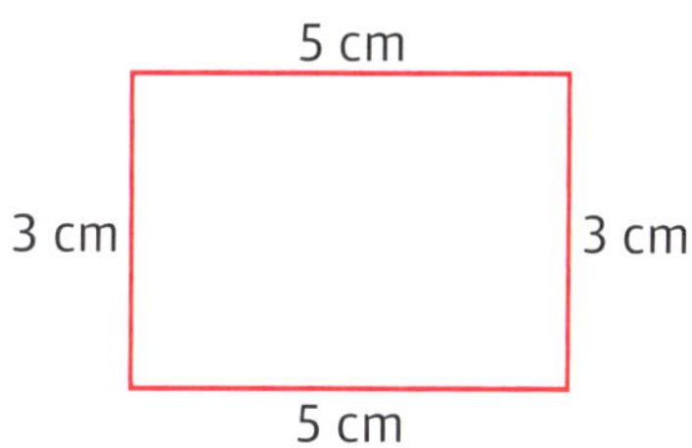

5 cm + 3 cm + 5 cm + 3 cm = 16 cm

The perimeter of this shape is 16 cm.

A quick way to work out the perimeter of a shape is to use multiplication. For example, for this rectangle you could double the length and add it to double the width.

perimeter = (2 × length) + (2 × width) $p = 2l + 2w$

$2l = 2 \times 5 = 10$ cm

$2w = 2 \times 3 = 6$ cm

so, perimeter = 16 cm

To find the perimeter of an irregular shape, find a starting point and add up the lengths of all sides until you get back to the starting point.

Area

The area of a 2D shape is the space inside the perimeter, which could be coloured in.

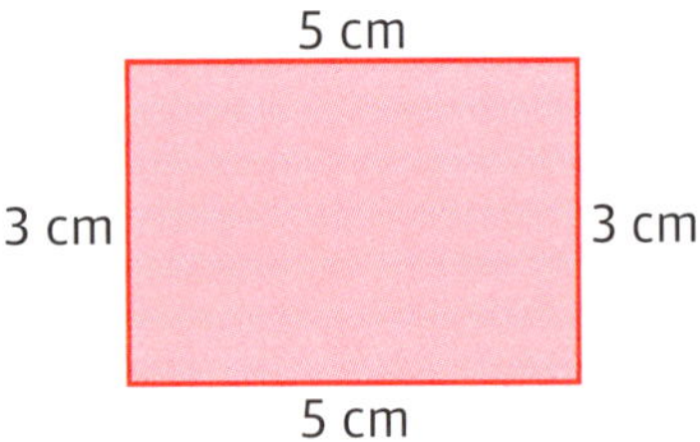

To find the area of this rectangle, multiply its length by its width.

$5\,\text{cm} \times 3\,\text{cm} = 15\,\text{cm}^2$

$A = l \times w$

The area of this shape is $15\,\text{cm}^2$.

To find the area of an irregular shape, you may have to split it up into smaller squares or rectangles. Find the area of each smaller shape and add them together.

You may have to count squares on a grid. Count the whole squares within
the shape followed by those squares where half or more than half of the square is included in the shape. The total is an approximate estimate of the shape's area.

LEARN: Area is a measurement of space covered. It might be measured in square centimetres (cm^2) for small areas, square metres (m^2) for large areas like football pitches, or square kilometres (km^2) for vast areas such as countries.

Look at this example:

What are the width and area of a rectangular field if the perimeter is 260 m and the length of the field is 90 m?

perimeter = 2 × length + 2 × width

260 m = 2 × 90 m + $2w$

260 m = 180 m + $2w$

260 m – 180 m = $2w$

80 m = $2w$

40 m = w

Width = 40 m

Area: length × width = 90 m × 40 m = 3600 m^2

90 m

90 m

You can use the same rule (or formula) to find the area of a parallelogram: base × height.

Be sure to remember to use the height, not the length of the other sides.

(To find the area of a triangle, see *Topic 18: 2D shapes: triangles*.)

vertical height

base

HAVE A GO

1 Find the area of a square that has a perimeter of 20 cm. ______

2 What is the perimeter of a field that has two sides of 45.5 m and two sides of 91.9 m? ______

3 How many tiles each measuring 20 cm × 20 cm would be needed to cover a rectangular work surface with measurements of 1 m × 60 cm? ______

4 A parallelogram has a length of 12 cm and a vertical height of 9 cm. What is its area? ______

5 The perimeter of a regular hexagon is 8.4 cm. Find the length of one side in mm. ______

EXAM TIP

You will often be given diagrams to work with for questions about area and perimeter in exams. If it is a paper test, it is fine to draw on it if you need to add or sort out information. If it is an online test, you will often have paper for doing workings, so you can draw your own diagrams. It is very important to see what is going on. Set out calculations step by step. Even if your final answer is incorrect, you may get points for showing your working.

KEY FACTS

- The perimeter of a 2D shape is the total distance round the edge of the shape and can be measured in millimetres (mm), centimetres (cm), metres (m) or kilometres (km).
- The area of a 2D shape is the space inside the perimeter, which could be coloured in. It can be measured in square centimetres (cm^2), square metres (m^2) or square kilometres (km^2).
- To calculate the perimeter or area of a shape, you need to know the distance along each side of the shape.

21 3D shapes

Naming 3D shapes

You need to recognise and be able to name these 3D shapes:

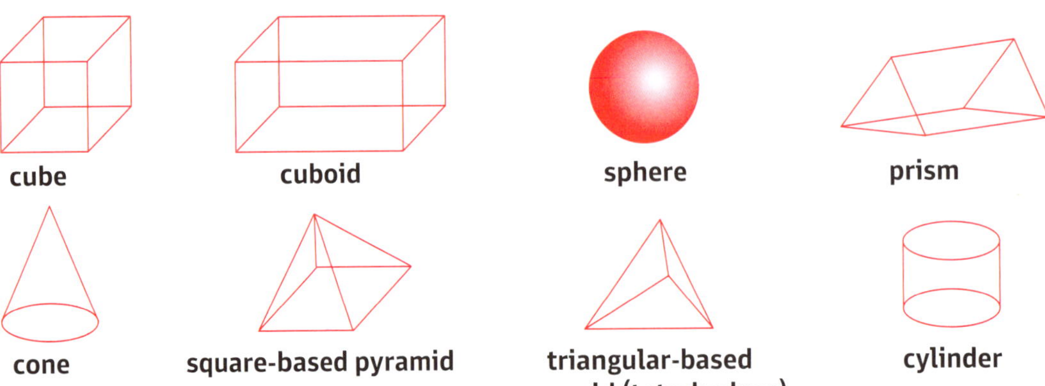

Prisms have the same shape at both ends with rectangles or squares joining the two ends together. Prisms are named according to their end shape.

Nets

A **net** is the shape you would draw, cut out and fold to make a 3D shape. To work out from a diagram whether a net will make a shape, you have to imagine cutting it out and folding it into the shape.

There are eleven ways of making the net of a cube. Here are two ways:

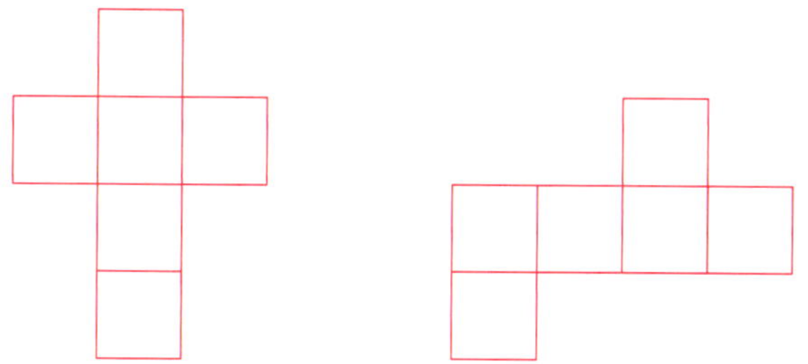

Faces, edges and vertices

- The **faces** of a solid 3D shape are the flat parts.
- The **edges** of a solid 3D shape are where two faces meet.
- The **vertices** of a solid 3D shape are the points or corners.

REMEMBER: The plural of vertex is vertices. A cone has one vertex. A cube has eight vertices.

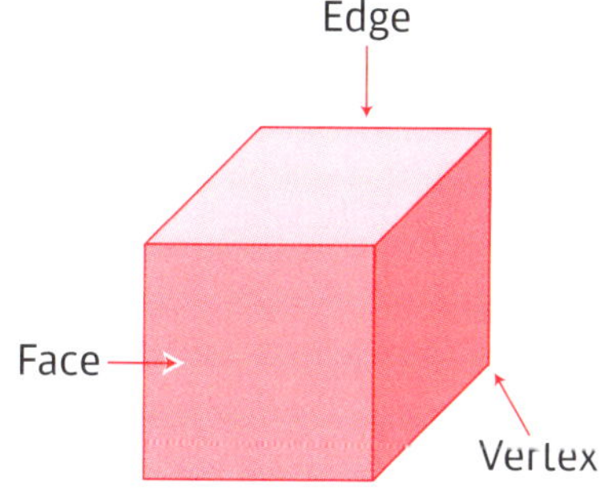

Shape and space

HAVE A GO

1 How many faces does a cuboid have? ______

2 How many vertices does a hexagonal prism have? ______

3 How many edges does a square-based pyramid have? ______

4 If you wanted to make a pentagonal prism from paper, what are the seven 2D shapes that would make up its net? ______

______

______

______

5 Find the two nets above for making cubes. There are nine more ways. Work out one way and try it out using paper.

KEY FACTS

- Prisms have the same shape at both ends with rectangles or squares joining the two ends together.
- A net is the shape you would draw, cut out and fold to make a 3D shape.
- The faces of a solid 3D shape are the flat parts.
- The edges of a solid 3D shape are where two faces meet.
- The vertices of a solid 3D shape are the points or corners.

22 *Volume and capacity*

LEARN: Volume is measured in cubic centimetres (cm^3) or cubic metres (m^3).

Volume

The **volume** of a solid 3D object is the amount of space it takes up. To find the volume of a **cube** or cuboid, multiply together the three dimensions: length, breadth and height.

Volume of a cuboid = $l \times b \times h$

The dimensions of this cuboid are:

$l = 4\,\text{cm}$

$b = 3\,\text{cm}$

$h = 2\,\text{cm}$

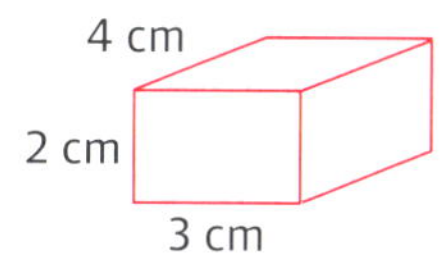

The volume of the cuboid is $4 \times 3 \times 2 = 24\,\text{cm}^3$.

To find the volume of an irregular solid object, split it up into smaller cubes or cuboids and find their separate volumes. Then add up or subtract the volumes.

Look at this irregular solid object.

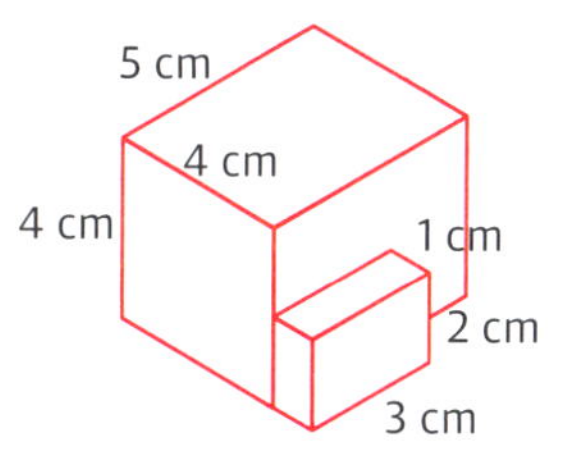

It can be split into two separate cuboids:

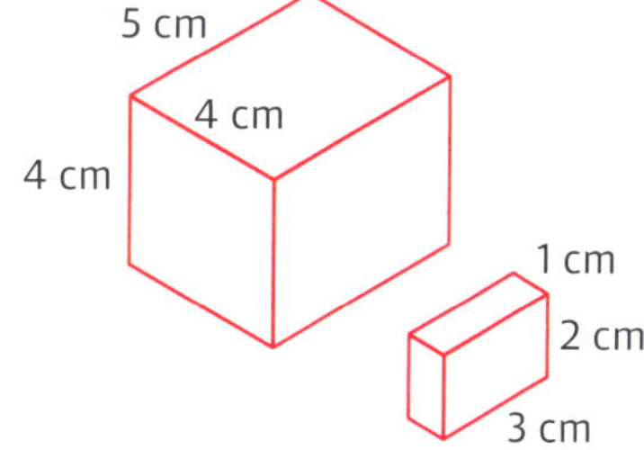

Large cuboid: $4 \times 4 \times 5 = 80\,\text{cm}^3$

Small cuboid: $2 \times 3 \times 1 = 6\,\text{cm}^3$

Total volume: $80 + 6 = 86\,\text{cm}^3$

Capacity

The **capacity** of a container is how much water or other liquid it will hold. Capacity can be measured in **litres** (l) and/or **millilitres** (ml).

LEARN:
1 litre = 1000 millilitres

The capacity 900 ml can also be written as 0.9 litres.

The capacity 1400 ml can also be written as 1 l 400 ml or 1.4 l.

One litre is just under two pints and is the amount in a standard carton of juice. It will give you four large glasses of juice.

A teacup holds about 200 ml. A mug holds about 300 ml.

Shape and space

HAVE A GO

1 The volume of a box is 3600 cm^3. If its length is 30 cm and it is 12 cm wide,what is its height? ______

2 How many millilitres are there in 2.3 litres? ______

3 What is the volume of this box if one of the cuboids that is used to make it has a volume of 9 cm^3? ______

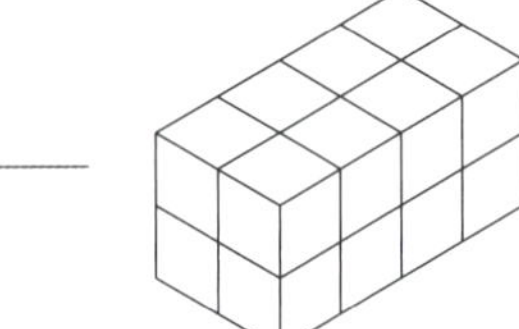

4 Three identical bottles each hold 650 ml. If they are filled up from a 2 litre bottle of orange juice, how much will be left in the larger bottle? ______

5 Granny's teapot holds just enough tea for four cups. Which is the most likely capacity of the teapot? Circle the correct answer.

2 litres   800 ml   300 ml   4 litres   0.6 litres

KEY FACTS

- Volume is the amount of space an object takes up and is measured in cubic centimetres (cm^3) or cubic metres (m^3).
- Volume of a cube/cuboid = length × breadth × height ($l \times b \times h$).
- Capacity is the amount of liquid a container will hold and is measured in litres (l) and/or millilitres (ml).

23 Coordinates and transformations

Coordinates

In questions about **transformations**, the corners (or **vertices**) of shapes are given as **coordinates**, such as (3,6). You may have to plot the shape on the ***x*- and *y*-axes** yourself.

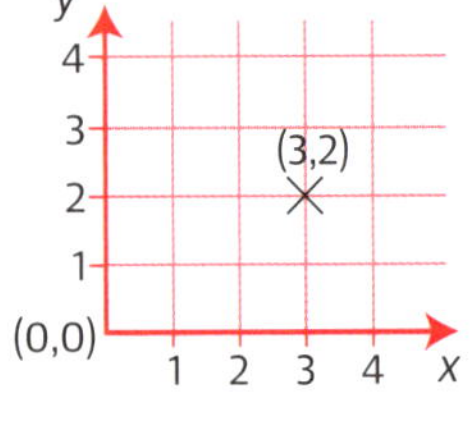

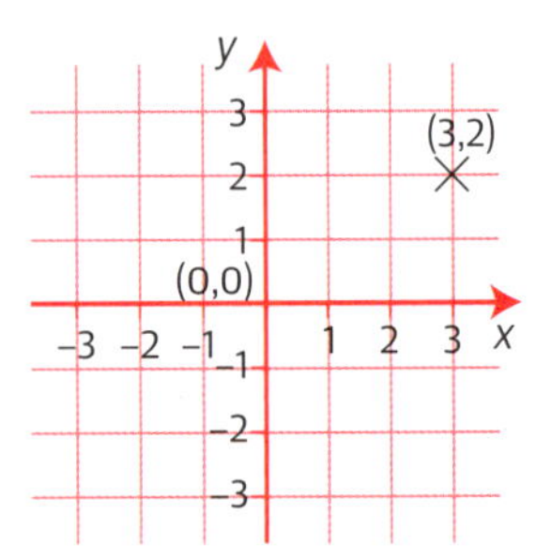

The first number in a pair of coordinates is always the *x* amount, the distance you have to go along the *x*-axis.

The second number in a pair of coordinates is always the y amount, the distance you have to go up or down the *y*-axis.

LEARN: Go 'along the hall first (*x*-axis) and then up or down the stairs (*y*-axis)' when giving a pair of coordinates.

Reflection, rotation and translation

There are three main types of transformation: **reflection**, **rotation** and **translation**.

- Reflection means the result of reflecting a shape in a given mirror line.
- Rotation means turning a shape round a given point or centre of rotation. An object can be rotated **clockwise**, like the hands of a clock, or **anticlockwise**, the opposite way from the hands of a clock.

clockwise  anticlockwise 

- Translation means moving a shape along and up or down.

Look at this example:

1 Plot these coordinates on the grid.

(–2,2), (–2,4), (–5,2)

Join them together (shape A).

2 Reflect the shape along the line $y = 1$ (shape B).

3 Rotate the original shape 180° clockwise about the point (0, 0) (shape C).

4 Translate shape C 1 square left and 4 squares down (shape D).

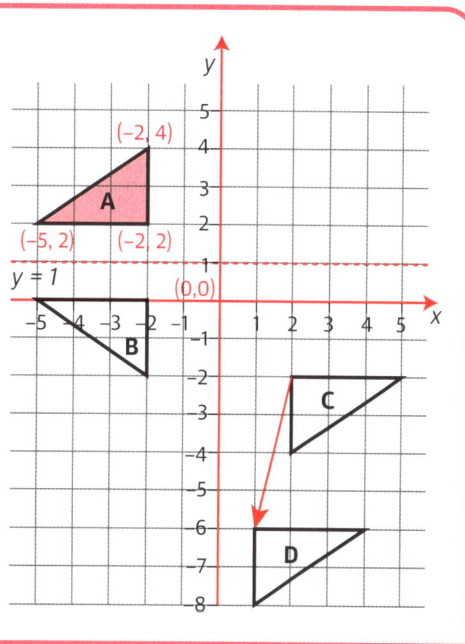

Shape and space

HAVE A GO

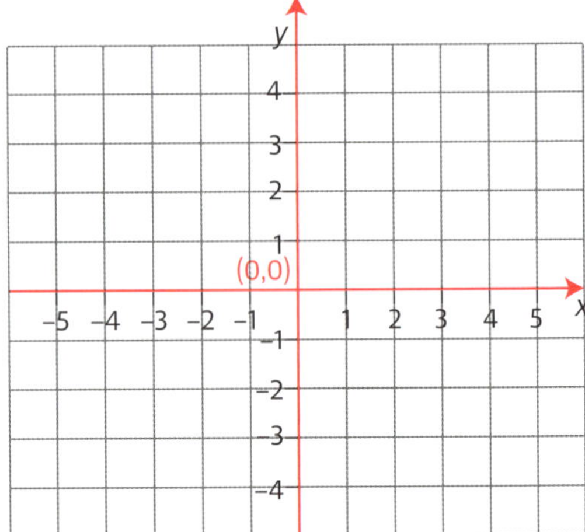

1 Plot these coordinates on the grid: (4,1), (5,1), (5,4).

Join them together and label the shape A.

2 Reflect the shape A along the line $x = 0$. Label it B.

3 Rotate shape A 90° anticlockwise about the point (3,1). Label it C.

4 Translate shape C 2 squares left and 5 squares down. Label it D.

5 Write down the coordinates of shape D. ______________________________

EXAM TIP

Be really careful to give coordinates starting with along the *x*-axis then up or down the *y*-axis. If you are rotating, you may be given tracing paper to use, but if not, try using a rough piece of paper and tracing the shape, thinking about the rotation you are asked to do. For translation, move each point in turn, following the instructions given.

KEY FACTS

- Coordinates tell you a specific point on a grid.
- The value on the *x*-axis is always given before the value on the *y*-axis: (x, y).
- Reflection is the result of reflecting a shape in a mirror line.
- Rotation means turning a shape round a point or centre of rotation.
- Clockwise is in the same direction as the hands of a clock and anticlockwise is the opposite direction to the hands of a clock.
- Translation means moving a shape along and/or up or down.

24 Symmetry

Line symmetry

A shape is said to have **line symmetry** (or to be **symmetrical**) when it can be divided into two identical, mirror images by a line of symmetry. Shapes can have more than one line of symmetry. Lines of symmetry are usually shown as dotted lines on a diagram.

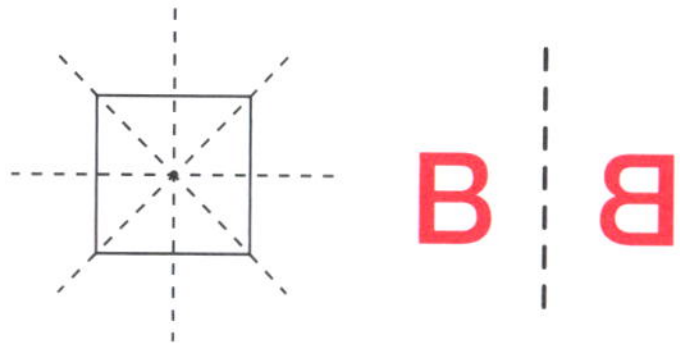

If you fold a shape along a line of symmetry, one side will fold onto the other.

REMEMBER: Folding is a good way to check if a line is a line of symmetry. The two sides must map (or fit) onto each other exactly.

Sometimes a line of symmetry is shown next to a shape and not on it at all. The shape is reflected symmetrically as if the line is a mirror. The distances between the points or **vertices** of the shape from the line of symmetry must be the same for both the shape and its reflected image:

Rotational symmetry

A shape has **rotational symmetry** if you can turn it around its centre and it maps (or fits) exactly on top of itself.

The number of times it does this is its **order of rotational symmetry**.

This star has rotational order of symmetry 5 because it maps onto itself five times before arriving back at where it started.

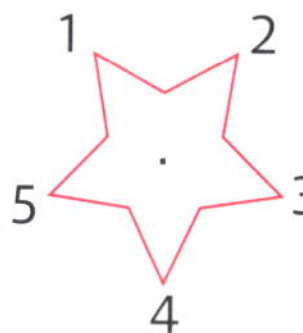

HAVE A GO

1 How many lines of symmetry does a regular hexagon have? ____________

2 What is the order of rotational symmetry of:

a an equilateral triangle ____________

b a rectangle? ____________

Shape and space

EXAM TIP

In an exam, remember that you can rotate your page or sketch out a quick diagram if it helps you. For symmetry questions, work point-to-point from wherever the mirror line is. The mirror line can be **vertical**, **horizontal** or **diagonal**.

KEY FACTS

- A shape is said to be symmetrical when it can be split into two identical, mirror images by a line of symmetry.
- Shapes can have more than one line of symmetry.
- A shape is described as having rotational symmetry if, when you turn it round, it maps onto itself exactly.
- The number of times a shape maps onto itself is its order of symmetry.

Measurement

25 *Metric and imperial units of measurement*

Metric measurements

All **metric units of measurement** use tens, hundreds and thousands and are simpler to use than **imperial** units.

Here are some details of metric units used to measure length, mass and **capacity**.

LEARN: milli → 1000th
deci → 10th
centi → 100th
kilo → 1000

LEARN: ≈ means 'approximately equal to'.

Metric equivalents	Examples
Length	
1 km (**kilometre**) = 1000 m (**metres**)	Kilometres are used to measure large distances between villages, towns and cities.
1 m = 100 cm (**centimetres**)	1 m is the length of a metre stick (look for one at school).
1 cm = 10 mm (**millimetres**)	1 cm is about the width of your fingertip. 1 mm is the width of a large full stop.
Mass	
1 t (**tonne**) = 1000 kg (**kilograms**)	1 tonne is about the weight of a small car.
1 kg = 1000 g (**grams**)	1 kg is the weight of a bag of sugar. 1 kg ≈ 2.2 lb
1 g = 1000 mg (**milligrams**)	1 g is the weight of a paper clip. Ingredients in medicines are often measured in milligrams.

Measurement

Metric equivalents	Examples
Capacity	
1l (**litre**) = 1000 ml (**millilitres**)	1l is the capacity of a standard fruit juice carton. 500 ml is the capacity of a standard washing-up liquid bottle. 330 ml is the capacity of a standard drink can. 5 ml is the capacity of a teaspoon.

It is particularly useful to be able to count in 25s up to 500 and in 250s up to 5000 to deal with parts of metric units of measurement. That way you can easily work out fractions of 100 and 1000, like $4\frac{3}{4}$ km = how many metres, or what is 3250 ml in litres?

Imperial measurements

Imperial units of measurement were used before metric ones replaced them. Imperial measurements continue to be used, particularly on road signs giving distances in **miles**. It is therefore important for you to know them and their equivalent metric amounts.

Imperial measurement	Approximate metric equivalents
Length	
1 mile = 1760 **yards**	Almost 2 km. 8 km ≈ 5 miles
1 yard (yd) = 3 feet	Just under 1 m.
1 **foot** (ft) = 12 **inches** 1 inch ≈ 2.5 cm	The length of a 30 cm ruler. The height of a tall man is about 6 ft or just under 2 m.

Imperial measurement	Approximate metric equivalents
Mass	
1 **stone** = 14 **pounds**	Just over 6 kg.
1 pound (lb) = 16 **ounces** (oz)	The amount of honey in a jar – almost 500 g.
1 ounce (oz)	The weight of three £1 coins. 1 oz ≈ 30 g
Capacity	
1 **gallon** = 8 **pints**	1 gallon ≈ 4 litres
1 pint	1 pint ≈ $\frac{1}{2}$ litre Milk is sometimes sold in pints.

It is important for you to know what these metric and imperial measurements look or feel like.

HAVE A GO

1 Complete: $3\frac{3}{4}$ m = ______________ cm

2 Which is heavier, 12 kg or 20 lb of potatoes? ______________

3 Complete the sentence.
A jug holding $2\frac{1}{2}$ litres would hold about ______________ pints of water.

4 If Jaz weighs 5 stone, how many kilograms is that? ______________

5 Circle the correct answers.

a 3 inches is about the same as: 10 cm 30 mm 250 mm 80 mm 0.3 m

b 16 km is about the same as: 16 miles 5 miles 160 miles 20 miles 10 miles

EXAM TIP

In an exam, make sure that you know the number of millilitres in a litre, grams in a kilogram and metres in a kilometre. It is also helpful to have a picture in your mind of examples of the different measurements. For example, a drink can is often 330 ml and the height of an average door is just under 2 m.

KEY FACTS

- There are metric and imperial units of measurement.
- Metric units of measurement are all based on tens, hundreds and thousands:
  1 km = 1000 m
  1 m = 100 cm
  1 cm = 10 mm
  1 tonne = 1000 kg
  1 kg = 1000 g
  1 g = 1000 mg
  1 litre = 1000 ml
- Imperial units of measurement include miles, stones and pints:
  1 mile = 1760 yards
  1 yard (yd) = 3 feet
  1 foot (ft) = 12 inches
  1 stone = 14 pounds
  1 pound (lb) = 16 ounces (oz)
  1 gallon = 8 pints
- These approximate equivalences show the relationships between metric and imperial measures:
  1 mile ≈ 2 km
  1 yard ≈ 1 metre
  1 kg ≈ 2.2 lb
  1 mile ≈ 2 km
  1 yard ≈ 1 metre
  1 ounce (oz) ≈ 30 grams (g) or 0.03 kilograms (kg)
  1 pint ≈ $\frac{1}{2}$ litre (l) or 500 millilitres (ml) or 0.500 l

26 Reading scales

Types of scales

A scale can be read in a straight line, like a ruler, or round, like a dial.

It is very important to look carefully for the **unit of measurement** given on a scale.

The jug says 1 litre at the top. The markings below it represent 50 **millilitres** and 100 **millilitres**.

The round dial says kg, meaning **kilograms**; therefore the smaller divisions each represent 100 **grams**.

REMEMBER: 1 l = 1000 ml
850 ml = 0.850 l
1 kg = 1000 g
28 g = 0.028 kg

It is important to work out what one division on a scale is measuring.

Here are some 1 litre jugs, each with different division lines marked on them.

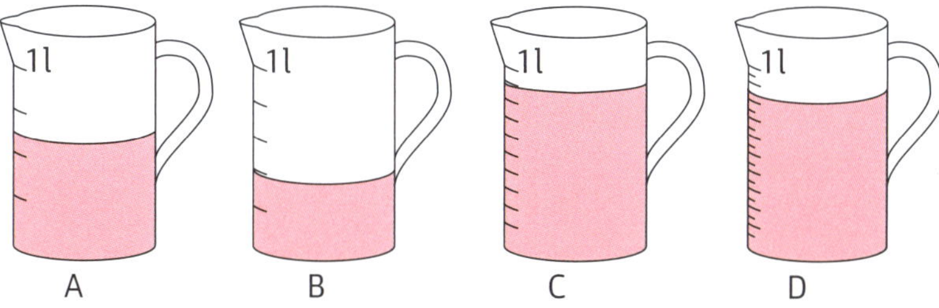

A 1000 ml ÷ 4 equal siz e sections = 250 ml per section.

B 1000 ml ÷ 5 equal size sections = 200 ml per section.

C 1000 ml ÷ 10 equal size sections = 100 ml per section.

D 1000 ml ÷ 20 equal size sections = 50 ml per section.

You also need to be able to measure **angles** by reading the scale on a **protractor**.

- Line up the bottom line of the angle with the base line of the protractor.
- Make sure the point of the angle is at the centre of the cross in the middle of the protractor.
- Find the scale that starts at 0 **degrees** for your angle. In the example below you want the inner scale.
- Read off the size of the angle using the correct scale. This angle is 45°.

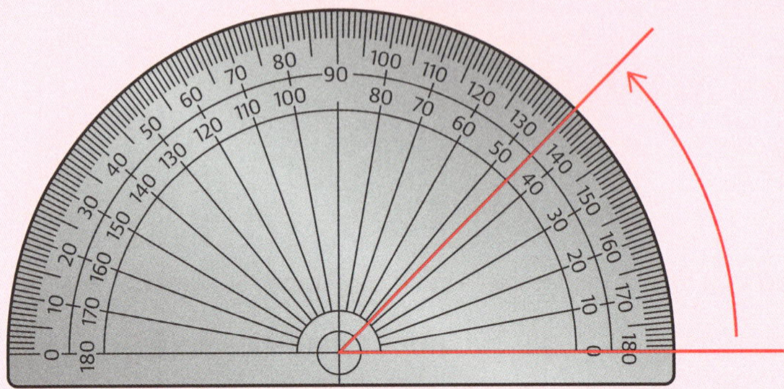

HAVE A GO

1 Work out how much liquid there is in each of the jugs A, B, C and D (on page 89) in ml.

A = ______________________ B = ______________________

C = ______________________ D = ______________________

2 a What is the mass shown by pointer A in kilograms and grams?

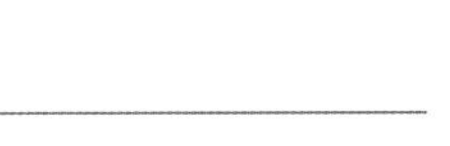

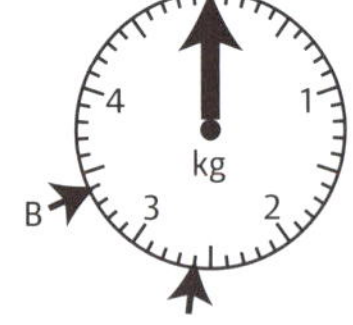

b How many grams must be added to the amount marked by pointer B to make 4 kg? ______________________

3 Use a protractor to measure these angles:

a

b

c

______________________ ______________________ ______________________

KEY FACTS

- A scale can be read in a straight line, like a ruler, or round, like a dial.
- Look for the units of measurement marked on a scale or dial.
- Calculate what one division is measuring on any scale you are given.
- When using a protractor to measure or draw angles, make sure you are using the right scale starting from zero, **clockwise** or **anticlockwise**.

27 Time and timetables

Telling the time

Telling the time is an essential skill. You need to be able to tell the time in words using a clock with hands (**analogue time**) and write a time in figures (**digital time**).

The 12-hour clock is based on the day having two sets of 12 hours. To make the difference clear between the times, **a.m.** (morning) or **p.m.** (afternoon) is written after the time. So, 9.00 a.m. is 9 o'clock in the morning and 9.00 p.m. is 9 o'clock in the evening.

The **24-hour clock** is shown on digital clocks. It continues after 12.00 at lunchtime like this: 13.00, 14.00 and so on, with 13.00 being 1 o'clock and 14.00 being 2 o'clock. So, 16.00 is 4 o'clock in the afternoon.

24-hour clock times always have four **digits** and therefore have a '0' written before the hours 1 to 9 in the morning: 03.25.

LEARN: Midnight in the 24-hour clock is zero hours: 00.00

Months

It is important to know how many days are in each month of the year.

This old rhyme can help you to remember how long each month lasts:

30 days has September,
April, June and November.
All the rest have 31
Excepting February alone,
Which has 28 days clear
And 29 in each **leap year**.

To adjust the calendar, a leap year comes every four years when the last two digits of the year are divisible by 4, except for the turn-of-the century years when the year has to be divisible by 400! The Summer Olympic Games usually happen in leap years.

Timetables

Timetables generally use the 24-hour clock.

To find the difference between two times, it can be useful to count the whole hours first and then count the minutes.

Look at this example:

If a bus leaves at 09:16 and arrives at its destination at 16:03, how long did the journey take?

09:16 → 15:16 = 6 hours

15:16 → 16:03 = 47 minutes

The journey took 6 hours 47 minutes.

HAVE A GO

1 How many hours are there in 2 and a half days? ______

2 How many minutes are there from 11:37 p.m. on Thursday until 1:13 a.m. on Friday? ______

3 The Masons' holiday is from 16 January until 13 February, including both dates. How many days is this? ______

4 If a train leaves at 10:27 and arrives at its destination at 15:09, how long does the journey take? ______

5 Fill in the gap: 40 mins + 38 mins + ______ mins = 2 hours

EXAM TIP

Telling the time and using timetables often comes up in exams. There will usually be a clock to look at in the exam room, so use this to help you work out times. Be sure you can tell the time in both analogue time using a.m. and p.m. and 12-hour and 24-hour digital time. It takes practice.

KEY FACTS

- The 12-hour clock uses the hours 1 to 12 twice in one day. Use a.m. to show a time in the morning and p.m. to show a time in the afternoon or evening.
- The 24-hour clock uses the hours 00.00 to 24.00, with 00.00 being midnight and 12.00 being midday. You do not use a.m. and p.m. with 24-hour times.
- Know how many days there are in each month and the order of the months of the year.
- Timetables generally use the 24-hour clock.

Skills builder

Try some of these activities to help improve your maths 11+ skills, some at home, some out and about.

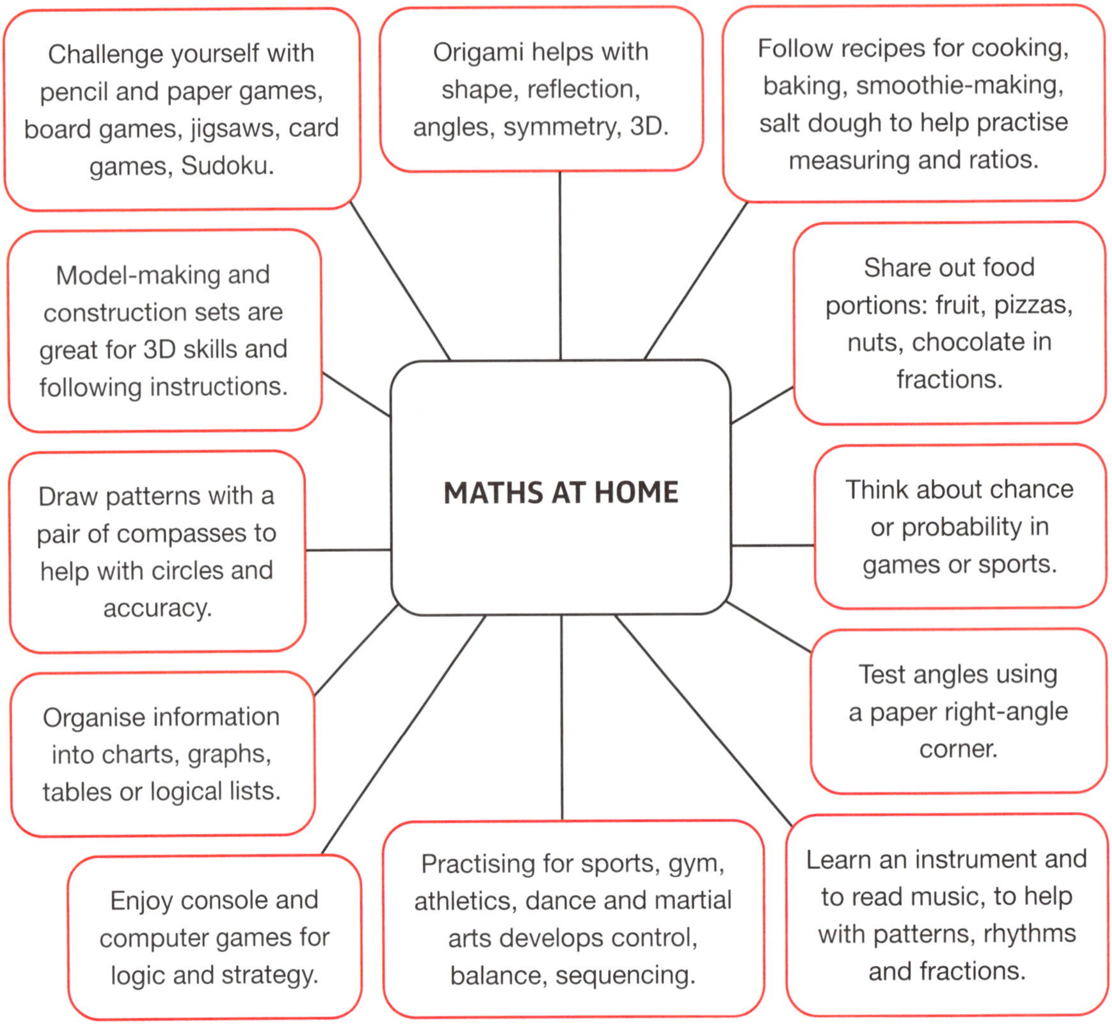

MATHS AT HOME

- Challenge yourself with pencil and paper games, board games, jigsaws, card games, Sudoku.
- Origami helps with shape, reflection, angles, symmetry, 3D.
- Follow recipes for cooking, baking, smoothie-making, salt dough to help practise measuring and ratios.
- Share out food portions: fruit, pizzas, nuts, chocolate in fractions.
- Think about chance or probability in games or sports.
- Test angles using a paper right-angle corner.
- Learn an instrument and to read music, to help with patterns, rhythms and fractions.
- Practising for sports, gym, athletics, dance and martial arts develops control, balance, sequencing.
- Enjoy console and computer games for logic and strategy.
- Organise information into charts, graphs, tables or logical lists.
- Draw patterns with a pair of compasses to help with circles and accuracy.
- Model-making and construction sets are great for 3D skills and following instructions.

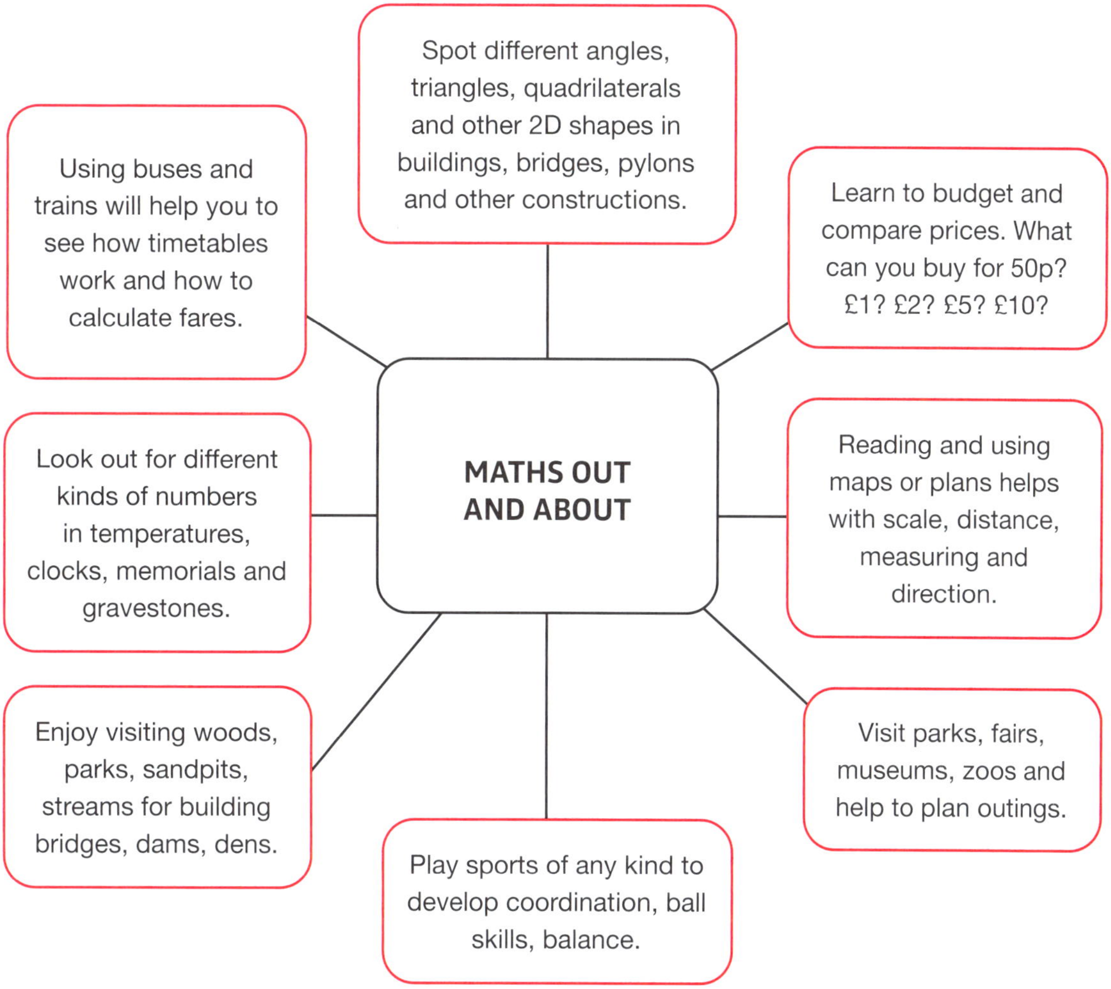

Skills builder

MATHS OUT AND ABOUT

Spot different angles, triangles, quadrilaterals and other 2D shapes in buildings, bridges, pylons and other constructions.

Learn to budget and compare prices. What can you buy for 50p? £1? £2? £5? £10?

Reading and using maps or plans helps with scale, distance, measuring and direction.

Visit parks, fairs, museums, zoos and help to plan outings.

Play sports of any kind to develop coordination, ball skills, balance.

Enjoy visiting woods, parks, sandpits, streams for building bridges, dams, dens.

Look out for different kinds of numbers in temperatures, clocks, memorials and gravestones.

Using buses and trains will help you to see how timetables work and how to calculate fares.

LEARNING AND PRACTISING MATHS SKILLS

- Spot patterns in flowers, leaves, tiles, fabrics.
- See how much things in the kitchen or bathroom measure or weigh.
- Use the sun or shadows or a compass to work out directions.
- Read big numbers in newspapers, or percentages in sales, or decimals in banks.
- Tell the time using analogue clocks with hands and see how this relates to digital time.
- Help with shopping, comparing prices, finding change.
- Open a savings account with pocket money or presents.
- Pack, wrap, weigh and work out postage for sending parcels.
- Brush up on maths 11+ skills before the exam.
- Follow instructions for origami, crafts, knitting, sewing.
- Learn to use the items in a simple geometry set.
- Share out money, nuts, chocolate, fruit, cake.

Study guide

You've worked through this book. Now test yourself!

1. Build confidence with practice

For more practice, and to put your skills to the test, work through the range of books and test papers in the Bond 11+ maths range.

Mark your answers with an adult. Talk about the questions you got wrong or found hard to understand. Read the sections in this book again to help brush up on things you are still not sure about.

It is a good idea to go over examples of things that might come up in an 11+ maths exam well before the date. A useful way to do this is to try some maths tests that are similar to the exam you will be doing. The Bond Assessment Practice books and test papers, when used regularly, will provide useful, graded practice that will build your confidence and show you how well you cope with doing tests of this kind.

2. Time yourself

If you are just starting to prepare yourself, you may find it helpful to go through your first few Bond Assessment Papers in maths untimed. This will help you to familiarise yourself with the types of questions and tasks you will face in the exam. Note down your scores (you could use the progress grids at the back of the Bond Assessment Practice books) and be sure to go over all the questions you found difficult until you understand them. After that, it is very important to give yourself a set time, just as you will have in the exam, so that you can practise pacing yourself and aiming to complete everything in the given time.

3. Revise strategies and techniques

It is also worth thinking about exam strategies and techniques. For many children, 11+ exams are the first exams they do in their lives, and they get very nervous at the thought of them. So do their parents! There are lots of hints on strategies and techniques in this book. Flick through them and talk about them to remind yourself. Highlight the ones that really work for you. It may be a good idea to make a reminder list of things to particularly look out for in the exam.

It is very important to remember that everyone is different. You will have your own way of coping and of doing things which may be quite different from the way other people work. If you have worked through this book, you will have a good idea of your own strengths and weaknesses, the things you find easy or difficult. You will have developed your own strategies and techniques in tests and in your learning.

4. Prepare for the exam day

"I'm so nervous..."

Of course you may be nervous, but actually many people find they can enjoy their exams if they feel confident and well prepared. After all, you will have done the practice; now it's your chance to show what you can do!

Just before the exam

Here are some useful things to remember before the exam day arrives and on the day itself:

- ✔ Don't worry about feeling a bit nervous; that's natural. Most children will feel anxious. Talk about your feelings and try to relax.
- ✔ Plan something fun to do after the exam is over.
- ✔ Try to have a good night's sleep.
- ✔ Eat a healthy breakfast and have something to drink.
- ✔ Make sure you have what you need:

  pencil/eraser/ruler/sharpener/tissues/glasses/water bottle/ inhaler, etc.
- ✔ Get to the place where the exam is happening in plenty of time.
- ✔ Find out where the toilets are and go if you can before the exam starts.

In the exam room

There should be no distractions during the exam because everyone is in the same boat as you and there will be at least one adult making sure that everything runs smoothly. The adult will tell you when to start the exam and when to stop. It is also their job to keep an eye on everyone and ensure there is no cheating.

However, there will be some distractions no one can do anything about. It could be a new, strange environment. It's your first real, public exam – and everyone else's too. People have distracting habits, like rocking their chairs, dropping things, muttering, fidgeting, sniffing... Some people may have a cough or a cold. Someone may need to go to the toilet. A child may not realise you can't ask for help and put up their hand to ask a question. Someone may have finished ages before you and is staring out of the window. It can all be very distracting! What can you do? The best thing is to ignore everything in the exam room apart from the adult in charge, the clock, the times written on the board and the exam paper in front of you.

Here are some useful strategies and techniques to remember once you are in the exam room:

- ✔ Keep calm. If you get butterflies or feel anxious, sit up straight, make sure your shoulders are not hunched and take some deep breaths. This allows plenty of oxygen to get to your brain, which needs it!
- ✔ Think positive. You've done all the hard work preparing. Now enjoy yourself!
- ✔ Find the clock. Make sure you know where it is before you start, so you can do a time check during the exam.
- ✔ Read the question. Not doing so is the most common mistake and easy to do something about.
- ✔ Write your answers carefully. Again, most mistakes are careless ones.
- ✔ Show what you know. This is your big moment and what you've practised for. Try to enjoy showing what you have learned.
- ✔ If you can't do a question, don't panic: have a go. Write something, and then put a mark in the margin, showing that you need to have another look if you have time at the end.
- ✔ Remember: a question left blank scores zero; a sensible guess may well be right.
- ✔ Leave time to check. Remember to leave a few minutes to check through your answers and make sure they make sense.
- ✔ Do your best: you can't do better than that!

GOOD LUCK!

Glossary

24-hour clock	the 24-hour clock uses the hours 00.00 to 24.00, with 00.00 being midnight and 12.00 being midday.
a.m.	the time between midnight and midday.
acute angle	an angle of less than 90°.
acute-angled triangle	this triangle has three acute angles.
adjacent	beside, near or next to.
algebra	the part of maths that deals with finding unknown numbers in equations.
analogue time	the time written or described in words using a clock with hands.
angle	measured in degrees, an angle tells us how far something turns or rotates.
anticlockwise	describing movement in the opposite direction to the way the hands of a clock move.
arc	a part of the circumference of a circle.
area	how much space there is inside a 2D shape, measured in square units, for example, cm^2, m^2, km^2.
average	a typical value of a set of data.
axis (plural: axes)	a graph has two lines called axes, the *x*-axis and *y*-axis, which join or intersect at the origin.
bearing	an angle between the direction north and the direction in which something is travelling.
cancelling	using division to find an equivalent fraction in its lowest terms.
capacity	how much liquid a container will hold, measured in litres (l) or millilitres (ml), or pints or gallons.
centimetre (cm)	one-hundredth of a metre.
circumference	the distance around the edge of a circle.
clockwise	describing movement in the same direction as the hands of a clock.
common denominator	a multiple of the denominators of two or more fractions; changing fractions to a common denominator allows you to compare, add and subtract the fractions.
common fraction	a fraction is a part of something OR a way to express a number less than one using numbers above (numerator) and below (denominator) a line.
complementary angle	two angles that together add up to 90°.
concentric circles	circles that share the same centre point but have a different radius.

cone	a 3D shape which has one flat face in the shape of a circle. The body of the shape is curved and leads up to a point or vertex (like an ice-cream cone).
consecutive numbers	numbers that follow on in order.
coordinates	these indicate the position of a point on a graph, for example, (3, 4). The first number is the distance you move in the *x*-direction; the second number is the distance you move in the *y*-direction.
cross-section	a drawing of something as if it has been cut through.
cube	a 3D shape with six identical square faces.
cube number	a number that is multiplied by itself twice.
cuboid	a 3D shape which has six rectangular faces.
cylinder	a 3D shape with a circle at each end and a curved surface joining them (like a tube or pipe).
data	collections of information.
decagon	a 2D shape with 10 straight sides.
decimal fraction	tenths, hundredths, thousandths, etc. shown as digits after a decimal point.
decimal number	a number written with a decimal point.
degrees	angles are measured in degrees. A small circle at the top right-hand side of the number is written to show that the measurement is in degrees. There are 360° in a full turn.
denominator	the bottom part of a fraction; it tells you how many equal parts the amount has been divided into.
diagonal	a diagonal joins two vertices of a polygon. A rectangle has two diagonals.
diameter	the distance straight across a circle from circumference to circumference, going through the centre.
digit	a single number from 0 to 9.
digital time	the time written in numbers, as you would see on a digital clock.
dividend	a number that is to be divided by another.
divisible	a number is divisible by a smaller number if the smaller number divides exactly into the larger number.
divisor	see dividend.
edge	the lines which join three or more vertices in a 2D shape or where two faces join in a 3D shape.
equation	a number sentence with an equals sign (=). The left-hand side has the same value as the right-hand side.
equilateral triangle	this triangle has three equal sides and three equal angles.
equivalent fractions	fractions that are equal to one another.

Glossary

estimate	make a sensible guess at an answer.
face	one surface of a solid 3D shape.
factor	a whole number that will divide exactly into another number.
foot (ft or ')	a measure of length, 12 inches (30.48 cm).
fraction	a part of something.
gallon	a unit used to measure liquids, equal to 8 pints or 4.546 litres.
gram (g)	a unit of mass or weight in the metric system.
graph	a visual way of displaying data; it can have bars, lines or pictures representing the data.
heptagon	a 2D shape with seven straight sides.
hexagon	a 2D shape with six straight sides.
highest common factor (HCF)	the HCF of a set of numbers is the largest number that is a factor of all numbers in the set.
horizontal	straight across.
hundredth	1 divided into 100 equal parts gives one hundred hundredths.
imperial	an imperial unit or measure is a non-metric one such as gallon, ounce or yard.
improper fraction	a top-heavy fraction where the numerator is larger than the denominator.
inch (")	a measure of length, one-twelfth of a foot (about 2.5 cm).
index (plural: indices)	the number of times a number is multiplied by itself. For example, $2 \times 2 \times 2 = 2^3$ has an index of 3.
integer	a whole number.
inverse	the opposite of something.
inverse operations	using the opposite of add, subtract, divide and multiply to find missing numbers.
irregular polygon	a 2D shape with sides that are not all the same length and the angles might not be the same.
isosceles triangle	this triangle has two equal sides and two equal angles.
kilogram (kg)	a unit of mass or weight equal to 1000 g (about 2.2 lb).
kilometre (km)	a unit of length equal to 1000 m (about $\frac{2}{3}$ of a mile).
kite	a four-sided polygon (2D shape with straight sides) which has two pairs of adjacent sides that are the same length.
leap year	a year with an extra day in it (29 February).
line symmetry	a shape has line symmetry when it can be divided into two identical mirror images. We say it is symmetrical.
litre (l)	a measure of liquid, about 1.75 pints.
lowest common multiple (LCM)	the LCM of a set of numbers is the smallest number that is a multiple of all numbers in the set.
lowest terms	a common fraction written using the smallest numbers possible.

mean	the number found by adding together all of the numbers in a set and dividing by how many numbers there are in the set.
median	the middle number in a set when the numbers are put in order of size.
metre (m)	a unit of length in the metric system, about 39.5 inches.
metric	to do with the metric system.
mile	a measure of distance equal to 1760 yards (about 1.6 km).
millilitre (ml)	one-thousandth of a litre.
mixed number	a mixture of a whole number and a fraction.
mnemonic	a verse or saying that helps you to remember something.
mode	the number in a set of data that occurs most often.
multiple	a multiple of a number is the answer when the number is multiplied by another number.
negative number	a number that is less than zero.
net	the shape you would draw, cut out and fold to make a 3D shape.
nonagon	a 2D shape with nine straight sides.
numerator	the top part of a fraction; it tells you how many equal parts you are interested in.
obtuse angle	an angle greater than 90° but less than 180°.
obtuse-angled triangle	this triangle has one obtuse angle.
octagon	a 2D shape with eight straight sides.
order of rotational symmetry	the number of times a shape maps exactly onto itself when turned around its centre.
origin	the point (0,0) on a graph.
ounce (oz)	a unit of weight equal to $\frac{1}{16}$ of a pound (about 28 g).
p.m.	the time between midday and midnight.
parallel	lines that are parallel never meet but run the same distance apart from each other for their entire length; they are not necessarily straight or equal in length.
parallelogram	a 2D four-sided shape that has its opposite sides parallel to each other.
pentagon	a 2D shape with five straight sides.
percentage	out of 100.
perimeter	the total distance around the edge of a 2D shape.
perpendicular	two lines are perpendicular if they are at right angles to each other.
pie chart	a circle divided into sections to show the proportions of how something is shared into groups.
pint	a measure for liquids, equal to one-eighth of a gallon (or 0.57 of a litre).
polygon	a 2D shape with three or more straight sides.
positive number	a number that is more than zero.
pound (lb)	a unit of weight, equal to 16 oz or about 454 g.

prime factor	the prime factors of a number are the prime numbers which can be multiplied together to make that number.
prime number	a prime number has only two factors: one and the number itself.
prism	a 3D shape with the same shape at each end and rectangles or squares joining the two ends together.
probability	the chance or possibility of something happening, usually written as a fraction.
profit	the extra money obtained by selling something for more than it cost to buy or make.
proportion	the fraction of the total amount when you divide up that amount using a given ratio.
protractor	a device for measuring angles, usually a semicircle marked off in degrees.
quadrilateral	a four-sided, 2D shape.
quotient	the number you get when you divide one number by another.
radius (plural: radii)	the distance from the centre of a circle to the circumference; half the diameter.
range	the range of a set of numbers is the difference between the smallest number and the largest number.
ratio	a ratio is used to compare two or more numbers or quantities and is usually expressed with a colon between the numbers.
rectangle	a 2D shape that has four straight sides and four right angles where the opposite sides are equal in length.
reflection	the result of reflecting a shape in a given mirror line.
reflex angle	an angle greater than 180° and less than 360°.
regular polygon	a 2D shape with sides the same length and with all angles equal.
remainder	the amount left over when one number does not divide exactly into another.
rhombus	a 2D shape with four equal sides where the sides are also parallel to each other; it looks like a slanting square or diamond shape.
right angle	an angle of 90° (think of the corner of a piece of paper).
right-angled triangle	this triangle has one right angle.
Roman numerals	capital letters, once used by Romans, which represent numbers.
rotation	turning a shape around a given point or centre of rotation, either clockwise or anticlockwise.
rotational symmetry	a shape has rotational symmetry if when you turn it around its centre it maps onto, or fits exactly on top of, itself.
scale factor	a number by which each of a set of quantities is multiplied.
scalene triangle	this has no equal sides and no equal angles.

sector	part of a circle between two radii, like a slice of pizza; semicircles and quadrants are sectors.
semicircle	half a circle.
simplest terms	see lowest terms.
sphere	a 3D version of a circle: a ball.
square	a regular polygon where the four sides and angles are the same size.
square number	the result of multiplying a number by itself.
square root	the square root of a number is the number you multiply by itself to make that number.
square-based pyramid	a 3D shape with a square base whose other edges meet at a point.
stone	a unit of weight, equal to 14 lb (6.35 kg).
symmetrical	see line symmetry.
tenth	1 divided into 10 equal parts give ten tenths.
tetrahedron	see triangular-based pyramid.
thousandth	1 divided into 1000 equal parts give a thousand thousandths.
tonne	a metric ton (1000 kg).
transformation	a way of moving a shape or an object, for example, rotation (turn), translation (slide) or reflection (flip).
translation	moving a shape right or left and/or up or down.
trapezium	a 2D shape that has four sides where one pair of sides is parallel but the other is not.
triangular number	a number that can be arranged as a triangle using, for example, dots.
triangular-based pyramid	a 3D shape with a triangular base whose other edges meet at a point.
units of measurement	a standardised way of showing the size of something, for example, litres.
value	the value of something is what it is worth or the amount represented by a letter or symbol.
Venn diagram	a diagram in which data is organised into overlapping rings.
vertex (plural: vertices)	the point or corner made where two straight lines join or where three or more edges of a solid 3D shape meet.
vertical	straight up or down.
volume	the amount of space a solid 3D object takes up, measured in cm^3 or m^3.
whole number	a number that does not have any fraction part to it. The counting numbers are whole numbers.
***x*-axis**	a graph's horizontal line.
yard (yd)	a measure of length, 36 inches or about 91 cm.
***y*-axis**	a graph's vertical line.

Answers

Answers

1 Place value

1 a 345 680 Look at the ones digit. 8 is more than 5, so round up.
b 345 700 Look at the tens digit. 7 is more than 5, so round up.
c 346 000 Look at the hundreds digit. 6 is more than 5, so round up.
d 350 000 Look at the thousands digit. It is 5, so round up.

2 $x = 30$ 3 tens = 30
$y = \frac{3}{100}$ 3 hundredths = $\frac{3}{100}$

3 1.6 Look at the hundredths digit. It is 5, so round up.

4 0.305 Move the digits three places to the right.

5 a 476.5 Look at the hundredths digit. It is 2, so round down.
b 476.53 Look at the thousandths digit. It is 8, so round up.

2 Addition and subtraction problems

1 4751

1786	
+ 2965	
3000	Total the thousands
1600	Total the hundreds
140	Total the tens
11	Total the ones
4751	The grand total!

2 1179

1786 →(+14)→ 1800 →(+1100)→ 2900 →(+65)→ 2965

14 + 1100 + 65 = 1179

3 £10.49 Round £7.99 up to £8, £8 + £2.50 = £10.50, then take off the 1p.

4 £3.66 Round £1.99 up to £2, £5.65 – £2 = £3.65, then add on the 1p.

5 4645

356 →(+4)→ 360 →(+40)→ 400 →(+600)→ 1000 →(+4000)→ 5000 →(+1)→ 5001

4 + 40 + 600 + 4000 + 1 = 4645

3 Multiplication and division problems

1 £41.70 Either use column multiplication or round £6.95 to £7. £7 × 6 = £42. 5p × 6 = 30p. £42 – 30p = £41.70

2 7 List the multiples of 24: 24, 48, 72, 96, 120, 144, 168 – 168 is the 7th multiple of 24.

3 90

$$70\overline{)6\,3\,0\,0}\quad = 9\,0$$

4 12 12 ÷ 2 = 6, 12 ÷ 3 = 4, 12 ÷ 4 = 3

5 34 650

	1	1		
		6	3	0
	×		5	5
	3	1	5	0
3	1	5	0	0
3	**4**	**6**	**5**	**0**

4 Mixed or several-step problems

1 £3.65 £20.00 – £1.75 = £18.25 (cost for 5 m). £18.25 ÷ 5 = £3.65 (cost per metre)

2 23 57 – 11 = 46 (children in the other two classes). 46 ÷ 2 = 23 (children in each class)

3 £2.25 £2.40 + £2.40 + £2.95 = £7.75 (total cost). £10.00 – £7.75 = £2.25 (change)

4 £4.90 £1.70 – £1.35 = £0.35 (profit per toy). £0.35 × 14 = (total profit)

5 8 18 × 9 = 162 (stickers used). 170 – 162 = 8 (stickers left)

5 Factors and multiples

1 60 The first six multiples of 10: 10, 20, 30, 40, 50, **60**
The first six multiples of 12: 12, 24, 36, 48, **60**, 72
The first six multiples of 15: 15, 30, 45, **60**, 75, 90

2 7 and 11 7 × 11 = 77

3 1, 2, 3, 4, 6, 8, 9, 12, 18, 24, 36, 72
Find the numbers in pairs: 1 × 72 = 72, 2 × 36 = 72 and so on.

4 30 The first 10 multiples of 3: 3, 6, 9, 12, 15, 18, 21, 24, 27, **30**
The first 10 multiples of 5: 5, 10, 15, 20, 25, **30**, 35, 40, 45, 50
The first 10 multiples of 6: 6, 12, 18, 24, **30**, 36, 42, 48, 54, 60

5	8	The factors of 32 are: 1, 2, 4, **8**, 16, 32 The factors of 88 are: 1, 2, 4, **8**, 11, 22, 44, 88 The factors of 120 are: 1, 2, 3, 4, 6, **8**, 10, 12, 15, 20, 30, 40, 60, 120

6 Special numbers

1	34	$3^2 = 3 \times 3 = 9$ and $5^2 = 5 \times 5 = 25$. $9 + 25 = 34$
2	36	$\sqrt{81} = 9$ ($9 \times 9 = 81$), $2^2 = 2 \times 2 = 4$ and the prime number between 20 and 28 is 23. $9 + 4 + 23 = 36$. Check: $1 + 2 + 3 + 4 + 5 + 6 + 7 + 8 = 36$, so 36 is a triangular number.
3	13, 14, 15	$42 \div 3 = 14$. 13 is 1 less than 14 and 15 is 1 more than 14, so $13 + 14 + 15 = 42$.
4	41, 43, 47	Prime numbers have only two factors: 1 and the number itself.
5	116	$\sqrt{81} = 9$ ($9 \times 9 = 81$) and $5^3 = 5 \times 5 \times 5 = 125$. $125 - 9 = 116$

7 Sequences

1	82	Add consecutive numbers: $47 + 5 = 52$, $52 + 6 = 58$, $58 + 7 = 65$, $65 + 8 = 73$, $73 + 9 =$ **82**
2	81	Subtract consecutive numbers: $100 - 10 = 90$, $90 - 9 =$ **81**, $81 - 8 = 73$ and so on.
3	16	Double each term: $4 \times 2 = 8$, $8 \times 2 =$ **16**, $16 \times 2 = 32$ and so on.
4	1	The triangular numbers: **1**, $1 + 2 = 3$, $1 + 2 + 3 = 6$, $1 + 2 + 3 + 4 = 10$ and so on.
5	32	Look at alternate numbers starting from 1: 1, 3, 5, 7, 9. These are the odd numbers. Look at alternate numbers starting from 2: 2, 4, 8, 16, ___. The rule is double each term: $16 \times 2 = 32$

8 Equations and algebra

1	5	Work backwards: half of 50 is 25 and $25 = 5 \times 5$
2	7	Add 9 to both sides ($4a + 21 = 7a$), then subtract $4a$ from both sides ($21 = 3a$) and finally divide both sides by 3 ($7 = a$)
3	45	Multiply both sides by 9 ($z = 45$)
4	31	$4 + 6^2 - 9 = 4 + 36 - 9 = 40 - 9 = 31$
5	8	$65 - 9 = 56$ and $7 \times 8 = 56$

9 Function machines

1	48	Start with the output number and work backwards: $12 \times 4 = 48$
2	28	Work from left to right: $59 - 3 = 56$ and $56 \div 2 = 28$
3	30	Work from left to right: $3 + 2 = 5$ and $5 \times 6 = 30$
4	3	Start with the output number and work backwards: $34 - 4 = 30$ and $30 \div 10 = 3$

10 Fractions

Fractions of numbers

1	9	Divide by the denominator: $27 \div 3 = 9$
2	36	Divide by the denominator: $63 \div 7 = 9$. Multiply by the numerator: $9 \times 4 = 36$
3	9	Divide by the denominator: $24 \div 8 = 3$. Multiply by the numerator: $3 \times 3 = 9$
4	60	Divide by the denominator: $72 \div 6 = 12$. Multiply by the numerator: $12 \times 5 = 60$
5	28	Divide by the denominator: $48 \div 12 = 4$. Multiply by the numerator: $4 \times 7 = 28$

Improper fractions

1	$8\frac{3}{5}$	$43 \div 5 = 8$ r 3
2	$12\frac{5}{6}$	$77 \div 6 = 12$ r 5
3	$9\frac{1}{10}$	$91 \div 10 = 9$ r 1
4	$15\frac{3}{4}$	$64 \div 4 = 15$ r 3
5	$9\frac{5}{12}$	$113 \div 12 = 9$ r 5

Equivalent fractions

$\frac{2}{3} = \frac{16}{24}$	$2 \times 8 = 16$ and $3 \times 8 = 24$
$\frac{3}{8} = \frac{21}{56}$	$3 \times 7 = 21$ and $8 \times 7 = 56$
$\frac{35}{45} = \frac{7}{9}$	$35 \div 5 = 7$ and $45 \div 5 = 9$
$\frac{1}{4} = \frac{21}{84}$	$1 \times 21 = 21$ and $4 \times 21 = 84$
$\frac{20}{48} = \frac{5}{12}$	$20 \div 4 = 5$ and $48 \div 4 = 12$

Comparing and ordering fractions

1 $\frac{4}{9}, \frac{7}{8}, \frac{9}{10}$, 1 and $\frac{2}{3}, \frac{13}{4}$

$\frac{4}{9}$ is less than $\frac{1}{2}$, so must be the smallest.

$\frac{1}{8}$ is bigger than $\frac{1}{10}$, so $\frac{9}{10}$ is bigger than $\frac{7}{8}$.

$\frac{2}{3} = \frac{8}{12}$ and $\frac{3}{4} = \frac{9}{12}$, so $\frac{13}{4}$ is bigger than 1 and $\frac{2}{3}$.

2 **a** $\frac{7}{12} < \frac{15}{24}$ — $\frac{7}{12} = \frac{14}{24}$ and $14 < 15$

b $\frac{6}{7} > \frac{5}{8}$ — $\frac{6}{7} = \frac{48}{56}$ and $\frac{5}{8} = \frac{35}{56}$, $48 > 35$

c $\frac{48}{96} = \frac{17}{34}$ — $\frac{48}{96} = \frac{1}{2}$ and $\frac{17}{34} = \frac{1}{2}$

Adding and subtracting fractions

1 $\frac{7}{11}$ — The denominators are the same, so add the numerators: $4 + 3 = 7$

2 $1\frac{3}{8}$ — Change the mixed number to an improper fraction with the denominator 8: $2\frac{1}{4} = \frac{9}{4} = \frac{18}{8}$. Then subtract the numerators: $18 - 7 = 11 \rightarrow 1\frac{1}{8}$. Write the answer as a mixed number. $11 \div 8 = 1$ r 3.

3 $1\frac{1}{8}$ — Change $\frac{3}{4}$ so the fractions have a common denominator: $\frac{3}{4} = \frac{6}{8}$. Then add the numerators: $3 = 6 = 9$. $\frac{9}{8}$ is an improper fraction: $\frac{9}{8} = 1\frac{1}{8}$

4 $\frac{19}{30}$ — Change to fractions with a common denominator before subtracting: $\frac{5}{6} = \frac{25}{30}$ and $\frac{1}{5} = \frac{6}{30}$. $25 - 6 = 19$.

5 $8\frac{3}{4}$ — Change to improper fractions with a common denominator before adding: $3\frac{1}{2} = \frac{14}{4}$ and $5\frac{1}{4} = \frac{21}{4}$. $14 + 21 = 35$. $\frac{35}{4}$ is an improper fraction: $\frac{35}{4} = 8\frac{3}{4}$

Simplifying fractions

1 $\frac{1}{6}$ — 5 and 30 have a common factor of 5. $5 \div 5 = 1$, $30 \div 5 = 6$

2 $\frac{1}{2}$ — 48 and 96 have a common factor of 48. $48 \div 48 = 1$, $96 \div 48 = 2$

3 $\frac{7}{10}$ — 70 and 100 have a common factor of 10. $70 \div 10 = 7$, $100 \div 10 = 10$

4 $\frac{6}{7}$ — 42 and 49 have a common factor of 7. $42 \div 7 = 6$, $49 \div 7 = 7$

5 $\frac{4}{5}$ — 32 and 40 have a common factor of 8. $32 \div 8 = 4$, $40 \div 8 = 5$

Dividing fractions

1 $1\frac{4}{5}$ — $\frac{3}{5} \times 3 = \frac{9}{5}$, $9 \div 5 = 1$ r 4

2 $2\frac{2}{9}$ — $\frac{5}{6} \times \frac{8}{3} = \frac{40}{18} = \frac{20}{9}$, $20 \div 9 = 2$ r 2

3 $\frac{1}{8}$ — $\frac{3}{4} \times \frac{1}{6} = \frac{3}{24} = \frac{1}{8}$

4 $2\frac{1}{10}$ — $\frac{7}{8} \times \frac{12}{5} = \frac{7}{2} \times \frac{3}{5} = \frac{21}{10} = 2$ r 1

5 $\frac{1}{14}$ — $\frac{2}{7} \times \frac{1}{4} = \frac{2}{28} = \frac{1}{14}$

11 Decimal numbers

Decimal fractions

1 **a** 0.8 — Try to learn some decimal fractions and their common fraction equivalents.

b 0.35 — $\frac{35}{100}$ = 3 tenths and 5 hundredths

c 0.678 — $\frac{678}{1000}$ = 6 tenths, 7 hundredths and 8 thousandths

d 0.35 — $7 \div 20 = 0.35$

e 0.125 — $1 \div 8$ 0.125

f 0.75 — Try to learn some decimal fractions and their common fraction equivalents.

2 **a** $5\frac{8}{10}$ or $5\frac{4}{5}$ — 5.8 has 5 ones and 8 tenths

b $11\frac{7}{10}$ — 11.7 has 10 tens, 1 one and 7 tenths

c $23\frac{6}{1000}$ or $23\frac{3}{500}$ — 23.006 has 2 tens, 3 ones, 0 tenths, 0 hundredths and 6 thousandths

d $7\frac{5}{10}$ or $7\frac{1}{2}$ — 7.5 has 7 ones and 5 tenths

e $4\frac{12}{100}$ of $4\frac{3}{25}$ — 4.12 has 4 ones, 1 tenth and 2 hundredths

3 5.032, 5.123, 5.14, 5.321, 5.322, 5.3, 5.33 — Look at each decimal place in turn. Remember that 5.3 = 5.300

4 0.008 — $0.2 \times 0.2 = 0.04$ and $0.04 \times 0.2 = 0.008$

5 a 0.27

$$\begin{array}{r} {}^{2}\,{}^{1} \\ 0.\not{3}0 \\ -\ 0.03 \\ \hline 0.27 \\ \hline \end{array}$$

b 35 There are 5 lots of 0.2 in 1, so there are 5 × 7 lots of 0.2 in 7

12 Percentages

1	£16.80	1% of 420 is 4.2, 4.2 × 4 = 16.8
2	45%	$0.45 = \frac{45}{100} = 45\%$
3	336	10% of 420 is 42. 20% is 2 × 42 = 84, so 84 children were absent. So 420 – 84 = 336 were present.
4	£29.75	10% of £35 = £3.50. 5% is half of £3.50 = £1.75. Saving is £3.50 + £1.75 = £5.25. Sale price is £35.00 – £5.25 = £29.75.
5	£15 080	10% of £13 000 is £1300. 5% is half of £1300 = £650. 1% is £130. Increase is £1300 + £650 + £130 = £2080. New price is £13 000 + £2080 = £15 080

13 Ratio and proportion

1	27	45 ÷ 5 = 9, 9 × 3 = 27
2 a	£2	1 + 2 + 5 + 8 = 16, so 1 share is £6.40 ÷ 16 = £0.40. Flynn's share is £0.40 × 5 = £2
b	$\frac{1}{8}$	Sophie's share is $\frac{2}{16} = \frac{1}{8}$
3	100 m	1 cm on the map represents 2000 cm = 20 m on the ground. 5cm on the map represents 5 × 20 m = 100 m on the ground.
4	39	4 + 3 = 7. 91 ÷ 7 = 13. 13 × 3 = 39.

14 Organising and comparing information

1	1	From the table, 12 children have cats. From the Venn diagram, 7 + 2 + 2 + ? have cats. So 7 + 2 + 2 + ? = 12, which gives 11 + ? = 12. So the missing number must be 1. Check using the number of dogs: 2 + 2 + 1 + 1 = 6 ✓ or Other pets: 4 + 2 + 1 + 1 = 8 ✓
2	30	$\frac{3}{8}$ of the children liked fish and chips best. $\frac{1}{8}$ of 80 is 10, so $\frac{3}{8}$ of 80 is 30.
3	13 km/h	Maryam travelled $6\frac{1}{2}$ km in 30 minutes. In an hour (2 × 30 mins) at the same speed, she would travel $6\frac{1}{2} \times 2 = 13$ km.
4	17	12 children belong to the Art group only and 5 children belong to neither group. 12 + 5 = 17.

15 Mean, median, mode and range

Range: 5	Largest value = 8, smallest value = 3. 8 – 3 = 5
Mode: 4	The mode is the value that appears most often. There are six 4s.
Median: 5	The median is the value in the middle when put in size order: 3, 3, 4, 4, 4, 4, 4, 4, **5**, **5**, 5, 5, 5, 6, 6, 6, 6, 6, 7, 8 There are 20 values. The 10th and 11th values are both 5.
Mean: 5	Find the total of the values and divide it by the number of values. 3 + 3 + 4 + 4 + 4 + 4 + 4 + 4 + 5 + 5 + 5 + 5 + 5 + 6 + 6 + 6 + 6 + 6 + 7 + 8 = 100. 100 ÷ 20 = 5.

16 Probability

1	5	A coin can land on heads or tails. So you would expect it to land on heads half the time.
2 a	$\frac{1}{6}$	There is only one way you can take out the ball numbered 5. There are six possible outcomes.
b	$\frac{3}{6}$ or $\frac{1}{2}$	There are three ways you can take out a ball that has an even number on it (2, 4, 6). There are six possible outcomes.
3	$\frac{5}{36}$	If you roll two dice, there are five ways the total can be 3 or 4 (1 + 2, 2 + 1, 1 + 3, 2 + 2, 3 + 1). There are 36 total possible scores.
4 a	$\frac{3}{7}$	There are 3 correct outcomes. There are 7 possible outcomes.
b	$\frac{4}{7}$	There are 4 correct outcomes. There are 7 possible outcomes.
c	0	There are 0 correct outcomes. It is impossible to pick a yellow ball.

5 $\frac{3}{13}$ There are 12 correct outcomes (3 royal cards in each of the 4 suits). There are 52 possible outcomes. $\frac{12}{52} = \frac{6}{26} = \frac{3}{13}$.

17 2D shapes: circles, angles and bearings

1 3.8 cm radius × 2 = diameter
2 North After 45° anticlockwise you are facing east. After another 90° anticlockwise you are facing north.
3 53° A right angle is 90°. 90 – 37 = 53.
4 260° Complementary angles add up to 360°. 360 – 100 = 260.
5 120° From 4 to 8 on a clock is $\frac{1}{3}$ of a whole turn.

18 2D shapes: triangles

1 55° The angles in a triangle add up to 180°. 80 + 45 = 125. 180 – 125 = 55.
2 B In an equilateral triangle, all three angles are the same – they are all 60°, which is an acute angle (less than 90°)
3 Scalene A scalene triangle has no equal sides and no equal angles.
4 42 cm² Area of a triangle = $\frac{1}{2}$ × base × height. $\frac{1}{2}$ × 12 × 7 = 42.

19 2D shapes: quadrilaterals and other polygons

1 50° The angles in a quadrilateral add up to 360°. 110 + 120 + 80 = 310. 360 – 310 = 50.
2 108° The interior angles of a regular pentagon add up to (5 – 2) × 180 = 3 × 180 = 540. One interior angle is 540 ÷ 5 = 108°.
3 6 A regular hexagon has six lines of symmetry.
4 130° The angles in a quadrilateral add up to 360°. Opposite angles in a parallelogram are equal. 50 + 50 = 100. 360 – 100 = 260. 260 ÷ 2 = 130.
5 Octagons and squares.

20 Perimeter and area

1 25 cm² The four sides of a square are the same length. 20 ÷ 4 = 5 cm. Area of a square = side length × side length. 5 × 5 = 25.
2 274.8 m 45.5 + 45.5 + 91.9 + 91.9 = 274.8 m
3 15 1 m = 100 cm. 100 cm ÷ 20 cm = 5. 60 cm ÷ 20 cm = 3. 5 × 3 = 15.
4 108 cm² Area of a parallelogram = base × height. 12 × 9 = 108.
5 16 mm A regular hexagon has six equal sides. 8.4 cm ÷ 6 = 1.6 cm. 1 cm = 10 mm. 1.6 × 10 = 16 mm.

21 3D shapes

1 6 Imagine a cuboid to help you.
2 12 A hexagon has six vertices. A hexagonal prism has a hexagon at each end.
3 8 The square base has four edges. Four more edges join the four vertices of the base to the top of the pyramid.
4 2 pentagons and 5 rectangles
A pentagonal prism has a pentagon at each end.
5 Check your net makes a cube.

22 Volume and capacity

1 10 cm Volume of a cuboid = length × breadth × height. 30 × 12 = 360. 3600 ÷ 360 = 10.
2 2300 ml There are 1000 millilitres in a litre. 2.3 × 1000 = 2300.
3 144 cm³ The box is made of 2 × 2 × 4 = 16 cuboids. 16 × 9 = 144
4 50 ml 650 × 3 = 1950. 2 litre = 2000 ml. 2000 – 1950 = 50.
5 800 ml A teacup holds about 200 ml. 200 × 4 = 800.

23 Coordinates and transformations

1–4

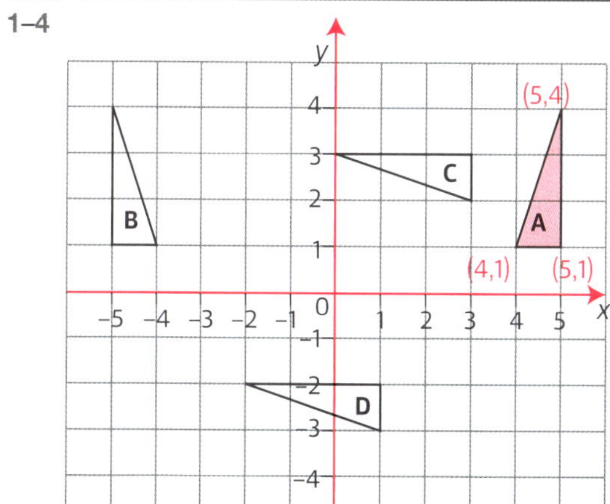

5 (–2, –2), (1, –2), (1, –3)

24 Symmetry

1	6	A regular hexagon has 6 equal sides and 6 lines of symmetry.
2 a	3	An equilateral triangle has 3 equal sides and 3 lines of symmetry.
b	2	A rectangle has two pairs of equal sides and 2 lines of symmetry.

25 Metric and imperial units of measurement

1	375 cm	3 m = 300 cm. $\frac{3}{4}$m = $\frac{3}{4} \times 100$ = 75 cm.
2	12 kg	6 kg is about 14 lb, so 12 kg is about 28 lb. 28 > 20.
3	5	$\frac{1}{2}$ a litre is about 1 pint. $2\frac{1}{2}$ litres is five lots of $\frac{1}{2}$ litre.
4	30 kg	1 stone is about 6 kg. 5 × 6 = 30.
5 a	80 mm	1 inch is about 25 mm. 3 × 25 = 75.
b	10 miles	8 km is about 5 miles. 5 × 2 = 10.

26 Reading scales

1	A = 625 ml	There are four divisions, so each division line represents 1000 ÷ 4 = 250 ml. The liquid is halfway between the second (500 ml) and third (750 ml) division lines.
	B = 400 ml	There are five divisions, so each division line represents 1000 ÷ 5 = 200 ml.
	C = 900 ml	There are 10 divisions, so each division line represents 1000 ÷ 10 = 100 ml.
	D = 850 ml	Each long division line represents 100 ml. There are two divisions between each pair of long division lines, so each short division line represents 100 ÷ 2 = 50 ml.
2 a	2 kg 600 g	For each kilogram, there are 10 divisions, so each division line represents 1000 ÷ 10 = 100 g.
b	600 g	Pointer B shows 3 kg 400 g. There are six more divisions before 4 kg.
3 a	105°	Make sure you line up your protractor correctly and read from the correct scale.
b	20°	
c	175°	

27 Time and timetables

1	60	There are 24 hours in one day. 24 + 24 + 12 = 60.
2	96 minutes	11:37 p.m. on Thursday to midnight is 23 minutes. Midnight to 1:13 a.m. on Friday is 60 + 13 = 73 minutes. 23 + 73 = 96.
3	29 days	There are 31 days in January. 16 to 31 January is 16 days. 16 + 13 = 29.
4	4 h 42 min	10:27 to 11:00 is 33 minutes. 11:00 to 15:00 is 4 hours. 15:00 to 15:09 is 9 minutes. 33 minutes + 4 hours + 9 minutes = 4 hours 42 minutes.
5	42 mins	40 mins + 38 mins = 78 mins = 1 hour 18 mins. 1 hour 18 mins + 42 mins = 2 hours.

Answers

Great Clarendon Street, Oxford, OX2 6DP, United Kingdom

Oxford University Press is a department of the University of Oxford.
It furthers the University's objective of excellence in research, scholarship, and education by publishing worldwide. Oxford is a registered trade mark of Oxford University Press in the UK and in certain other countries

Written by Liz Heesom

First published in 2024

British Library Cataloguing in Publication Data
Data available

ISBN: 9781382054171

10 9 8 7 6 5 4 3 2 1

Printed in the UK

The manufacturing process conforms to the environmental regulations of the country of origin

Acknowledgements

The Publishers would like to thank Michellejoy Hughes for her contribution to this edition.

Cover illustrations by Lo Cole
Typeset by Integra